DATE DUE

5/16/01			
GAYLORD			PRINTED IN U.S.A.

Biological Significance of Superantigens

Chemical Immunology

Vol. 55

Series Editors
Frank W. Fitch, Chicago, Ill.
Kimishige Ishizaka, La Jolla, Calif.
Peter J. Lachmann, Cambridge
Byron H. Waksman, New York, N.Y.

KARGER

Basel · Freiburg · Paris · London · New York · New Delhi · Bangkok · Singapore · Tokyo · Sydney

Biological Significance
of Superantigens

Volume Editor
Bernhard Fleischer, Mainz

23 figures and 14 tables, 1992

KARGER

Basel · Freiburg · Paris · London · New York · New Delhi · Bangkok · Singapore · Tokyo · Sydney

Chemical Immunology

Formerly published as 'Progress in Allergy'
Founded 1939 by Paul Kallòs

RC
583
.P7
V.55

Bibliographic Indices
This publication is listed in bibliographic services, including Current Contents® and Index Medicus.

Drug Dosage
The authors and the publisher have exerted every effort to ensure that drug selection and dosage set forth in this text are in accord with current recommendations and practice at the time of publication. However, in view of ongoing research, changes in government regulations, and the constant flow of information relating to drug therapy and drug reactions, the reader is urged to check the package insert for each drug for any change in indications and dosage and for added warnings and precautions. This is particularly important when the recommended agent is a new and/or infrequently employed drug.

Contents

Contents

Contents

Effects of Staphylococcal Toxins on T-Cell Activity in vivo
Ochi, A.; Yuh, K.; Migita, K.; Kawabe, Y. (Toronto, Ont.)

Mycoplasma arthritidis-Derived Superantigen
Rink, L.; Kirchner, H. (Lübeck)

Cellular and Molecular Mechanisms of Immune Activation by Microbial Superantigens: Studies Using Toxic Shock Syndrome Toxin-1
Chatila, T.; Scholl, P.; Ramesh, N.; Trede, N.; Fuleihan, R.; Morio, T.; Geha, R.S. (Boston, Mass.)

Contents

Introductory Remarks

'Superantigens' is the designation of a group of molecules that have in common an extremely potent activating effect on T lymphocytes of several species. This designation originates from the finding that the mechanism of T lymphocyte stimulation closely mimics the recognition of specific antigen: superantigens cross-link variable parts of the T-cell antigen receptor with MHC class II molecules on accessory or target cells. Toxic shock syndrome toxin 1 is the prototype superantigen. The superantigens known so far are produced by bacteria (*Staphylococcus aureus* and *Streptococcus pyogenes*, both gram-positive cocci), a mycoplasma (*Mycoplasma arthritidis*) and by murine retroviruses (mouse mammary tumor viruses). Thus, surprisingly distinct pathogens have evolved the same basic mechanism of T-cell stimulation in evolution.

The interaction site on the $\alpha\beta$ T-cell receptor is the variable part of the β-chain (V_β). Stimulation of T cells with a superantigen in vitro leads to a dramatic increase of T-cell subsets carrying certain V_βs. In vivo, the application of a superantigen leads to a transient expansion and subsequent death and anergy of T cells with the appropriate V_βs. That such selective enrichment or deletion occurs in certain immunopathological conditions, e.g. AIDS or rheumatoid arthritis, has been taken as evidence for the involvement of superantigens. This opens exciting aspects for the role of these molecules in human immunopathology. One has to keep in mind, however, that also antigenic peptides can selectively stimulate T cells carrying certain V_βs or V_γs. Due to the popularity of superantigens and probably due to their attractive name, a number of other candidate molecules have been proposed to belong to this group. For these, however, the definite proof of their superantigenicity is still missing.

Most contributions to this book describe the effects of superantigens on T lymphocytes. Although the basic mechanisms of T-cell stimulation by these molecules is similar, each superantigen has its own unique features. The enormous potency of these molecules is reflected by the biological consequences found when superantigens are introduced into the body. Most prominent and discussed in several contributions are the shock-like symptoms induced by the staphylococcal enterotoxins and the toxic shock syndrome toxin 1. It is now recognized that these symptoms are caused by an effect of the toxins on T cells, leading to a massive release of lymphokines and monokines.

Another important feature of all superantigens is the induction of immunosuppression in vivo by superantigens. Although the underlying molecular mechanisms are not completely clear, it is accepted that death of T cells occurs after an initial stimulation in vivo. In the long run this may open possibilities for the selective depletion of T cells for therapeutic purposes.

Progress in research on superantigens has been extremely rapid in the last few years and is accelerating. It is possible that as this book goes to print it will miss important developments. We can expect further exciting findings in the near future.

Fleischer B (ed): Biological Significance of Superantigens.
Chem Immunol. Basel, Karger, 1992, vol 55, pp 1–35

Staphylococcal Enterotoxins, Toxic Shock Syndrome Toxin and Streptococcal Pyrogenic Exotoxins: A Comparative Study of Their Molecular Biology

Marsha J. Betley, Deborah W. Borst, Laura B. Regassa

Department of Bacteriology, University of Wisconsin-Madison, Madison, Wisc., USA

Overview

This article reviews molecular biological and genetic aspects of the staphylococcal enterotoxins, toxic shock syndrome (TSS) toxins, and streptococcal pyrogenic exotoxins [reviewed in ref. 1–8]. Relationships among these toxins are defined based on nucleotide and amino acid sequence comparison data. The genetic elements containing the toxin genes and their genetic map locations are discussed. The section on gene expression reflects the fact that more work has been done on the staphylococcal toxins than the streptococcal pyrogenic toxins. Molecular biological and genetic aspects of the various genes are compared and contrasted.

Introduction

The staphylococcal enterotoxins, TSS toxins and streptococcal pyrogenic exotoxins, which will be collectively referred to as pyrogenic toxins, have common biological activities. They are pyrogenic, stimulate T cell proliferation, suppress immunoglobulin secretion, and enhance endotoxic shock and a hypersensitive skin reaction [reviewed in ref. 9–14]. Each of these toxins was recognized initially for its role in a specific disease.

Streptococcal pyrogenic exotoxins (also referred to as erythrogenic toxins, scarlatinal toxin, Dick toxin, scarlet fever toxin, cardiac toxins and erysipelas) are produced by some β-hemolytic *Streptococcus pyogenes* strains.

There are 3 antigenically distinct types (designated SPE A, SPE B and SPE C) [15]. Recently, variant alleles of *speA* have been described [16]. These cardiac toxins are well known for their role in inducing symptoms of scarlet fever [17].

Staphylococcal enterotoxins are the emetic toxins that cause staphylococcal food poisoning syndrome [9]. Five major serological types have been characterized (A–E; referred to as SEA, SEB, etc.) [9]. There are 3 minor subtypes of SEC, designated SEC1, SEC2 and SEC3, which differ by minor antigenic epitopes [9]. In addition to the functional enterotoxin genes, a nontranslated enterotoxin-like gene has been identified, $sezA^+$ [18]. $sezA^+$ has substantial nucleotide sequence identity with sea^+ and see^+ but is apparently not translated because it lacks a translation initiation signal [18]. For the staphylococcal enterotoxin genes, the superscript + indicates a wild-type allele; this convention is part of the guidelines for the genetic nomenclature for the staphylococcal enterotoxin genes [19].

TSS toxin 1 (designated TSST-1) from human isolates of *Staphylococcus aureus* was initially recognized as a major virulence factor in the human infection TSS. TSST-1 induction of high levels of cytokine secretion, such as tumor necrosis factor and interleukin-1, is thought to contribute to many of the shock manifestations [20–23]. Based on isoelectric point (pI) determinations, variant forms of TSST-1 have been identified from *S. aureus* strains associated with mastic disease in animals [24, 25].

The enterotoxins and streptococcal pyrogenic toxins have been implicated as virulence factors in toxic-shock-like syndromes. By analogy to TSST-1, some of the clinical symptoms are probably due to stimulation of cytokine secretion by immune system cells due to the enterotoxins and streptococcal pyrogenic toxins. Although most menstrual TSS cases are due to *S. aureus* TSST-1-producing strains ($\geq 90\%$), only about half of the nonmenstrual cases of TSS are due to strains that produce TSST-1 [5]. Many of the remaining nonmenstrual TSS isolates produce a staphylococcal enterotoxin, with SEB being the most common [26–31]. In recent years, severe forms of streptococcal disease have been described including some that resemble TSS; SPE A has been implicated as a causal factor in these toxic-shock-like cases [32–36].

Initial Characterization of Cloned Toxin Genes

The genes for the staphylococcal enterotoxins [37–44], TSST-1 [45] and the streptococcal pyrogenic toxins [46–49] have been cloned using *Escheri-*

chia coli host and vector systems. Verification of recombinant clones was based, at least in part, on serological detection of the toxin produced by *E. coli*. The promoter for each of these toxin genes, except *seb*⁺, is recognized apparently by *E. coli* RNA polymerase. Expression of *seb*⁺ is observed for an *E. coli* strain that has a transcriptional fusion between *seb*⁺ and the lambda P_R promoter [38]. Most of the work with recombinant toxin genes has been done using *E. coli* systems; in addition, stable toxin-producing *Bacillus subtilis* strains have been obtained by transfer of SPE-A and TSST-1-encoding plasmids [50].

Examination of nucleotide and amino acid sequence data is consistent with all of these toxins being produced in a precursor form [40–45, 51–61]. The sizes of the structural genes and mature proteins are given in table 1. Apparently, the signal sequence of the precursor forms is cleaved to release the extracellular forms. All of these toxins, except SPE B, undergo little or no additional processing. The mature form of SEC3 may not be the result of simply removing the precursor's signal sequence. The N-terminal residue of SEC3 produced by FRI913 has been reported to be a glutamic acid residue by one group and a serine residue by another group [44, 62]; this difference may result from the removal of 27- or 28-residue signal sequences, respectively. It seems more likely that extracellular SEC3 is released after cleavage of a 27-residue signal sequence and the N-terminal glutamic acid residue is subsequently removed in the culture or during protein purification.

Unlike the other pyrogenic toxins, the mature form of SPE B appears to be the result of extensive processing. SPE B apparently is a processed form of the streptococcal proteinase precursor (SPP). The first evidence of this relationship between SPE B and SPP was the observation that these 2 proteins are serologically identical in Ouchterlony assays using antiserum prepared against SPE B, although they have different molecular weights [63]. Data from Hauser and Schlievert [61] agree with this model. The amino acid sequence inferred from the nucleotide sequence of *speB* was compared to the N-terminal amino acid sequence of SPE B either purified in the presence of protease inhibitors or purified by the standard protocol [63]. These data are consistent with SPE B being synthesized in a precursor form of 398 amino acid residues. The precursor is processed yielding an exoprotein of 371 residues, which upon subsequent proteolysis yields a breakdown product of 253 residues resulting in the mature form of SPE B [61]. Comparison of the derived amino acid sequence of *speB* of strain 86-858 (a strain lacking SPP-like proteolytic activity) to the sequence of SPP obtained from a proteinase-producing streptococcal strain revealed differences in 13 locations. Some of these differences

Table 1. Size of staphylococcal and streptococcal pyrogenic toxins

Toxin	Number of amino acid residues	
	precursor form[1]	mature form[2] (MW)
Staphylococcal enterotoxins		
SEA[3]	257	233 (27,100)
SEB[4]	266	239 (28,336)
SEC1[5]	266	239 (27,531)
SEC2[6]	266	239 (27,589)[7]
SEC3[8]	266	239 (27,563)[9]
SED[10]	258	228 (26,360)
SEE[11]	257	230 (26,425)
SEG[12]	258	233 (27,107)[13]
Streptococcal pyrogenic exotoxins		
SPE A[14]	251	221 (25,787)
SPE B[15]	398	253 (27,588)
SPE C[16]	235	208 (24,354)
Toxic shock syndrome toxins		
TSST-1[17]	234	194 (22,049)
TSST-O[18]	234	194 (22,095)[19]

[1] Derived from nucleotide sequence data. [2] Determined by comparison of data from DNA and protein analyses. [3] Reported by Betley and Mekalanos [59]. [4] Reported by Jones and Khan [55]. [5] Amino acid sequence data derived from the nucleotide sequence of sec^+ of S. aureus strain MN Don (designated $sec^+_{MN Don}$) reported by Bohach and Schlievert [57]. [6] Nucleotide sequences of sec^+_{FRI361} and $sec^+_{FRI1230}$ reported by Bohach and Schlievert [42] and Couch and Betley [43], respectively, are identical. [7] Mature form of SEC2 reported by Bohach and Schlievert [42] was used which is 1 amino acid residue longer than that reported by Couch and Betley [43]. [8] Derived amino acid sequence data for sec^+_{FRI913} reported by Hovde et al. [44]. [9] Mature form of SEC3 of strain FRI913 reported by Hovde et al. [44] was used which is 1 amino acid residue longer than the mature form of SEC3 of strain FRI913 reported by Reiser et al. [62]. [10] Reported by Bayles and Iandolo [41]. [11] Reported by Couch et al. [40]. [12] Munson and Betley [unpubl. results]. [13] The derived amino acid sequence of seg^+ was used, assuming that the signal sequence cleavage site corresponds to that of SEB. [14] Derived amino acid sequence of SPE A, reported by Weeks and Ferretti [56], which is identical to that of speA1, reported by Nelson et al. [16]. [15] Reported by Hauser and Schlievert [61]. The precursor form of SPE B presumably has a 27-amino-acid signal sequence which is removed to release a 371-residue extracellular protein that undergoes additional processing. [16] Reported by Goshorn and Schlievert [60]. [17] Reported by Blomster-Hautamaa et al. [53]. [18] Kreiswirth and Schlievert [unpubl. results]. [19] The derived amino acid sequence of TSST-O gene was used, assuming that the signal sequence cleavage site corresponds to that of TSST-1.

were suggested to be due to difficulties with amino acid sequence analysis [61, 64, 65]. Based on these results, it has been suggested that there are 2 subtypes of SPE B, one with and the other without proteolytic activity [61].

The existence of additional serological types of enterotoxins was predicted by identification of *S. aureus* strains that produced emetic toxins that did not react with antiserum produced against the characterized types [66]. Recently, we have identified and partially characterized a new enterotoxin gene designated *seg+* [Munson and Betley, unpubl. data]. While examining DNA from strains that produced uncharacterized enterotoxins, a strain was identified that hybridized to an *sec+*-containing probe only under conditions of reduced stringency; DNA containing *seg+* was cloned. Analysis of staphylococcal recombinant clones that are isogenic except for *seg+* demonstrated that *seg+* does encode an emetic toxin, thereby fulfilling the definition of a staphylococcal enterotoxin. The nucleotide and derived amino acid sequences of *seg+* are included in the comparisons described below.

Sequence Comparisons

Relationships among the Toxins

Significant nucleotide and amino acid sequence relationships have been reported among the enterotoxins, SPE A and SPE C [40–44, 49, 51, 52, 54–60]. For this article, sequence comparisons for the pyrogenic toxins were performed to generate comparison data for each pair using the same algorithms. Optimized homology scores for the amino acid sequences of the mature forms and the percentage of amino acid sequence identity were obtained with the computer program FASTA [67] (tables 2, 3). The significance of each optimized homology score was judged using the program RDF2 [67] (table 3). Amino acid sequence alignments including most of these toxins have been published [4, 6, 14]. Nucleotide sequence comparisons were done using the FASTA programs (data not shown).

There is significant amino acid sequence homology among all the enterotoxins and between each of the enterotoxins and SPE A (table 2). SPE C is a distant member of this group. The optimized homology scores between SPE C and the enterotoxin-SPE A group are usually lower than those observed between members of the enterotoxin-SPE A group (table 2). Also, between SPE C and members of the enterotoxin-SPE A groups, there are lower percentages of amino acid sequence identities, and these identities occur over smaller regions as compared to that observed among other members of the enterotoxin-SPE A

Table 2. Optimized homology scores observed between amino acid sequences of mature forms of toxins and the maximum scores between the test and shuffled query sequence (given in parentheses)

Test sequence	Query sequence					
	SEA	SEB	SEC1	SEC2	SEC3	SED
SEA	1,160 (67)	149 (64)	150 (57)	150 (61)	150 (72)	709 (58)
SEB		1,211 (65)	485 (84)	485 (62)	465 (69)	132 (56)
SEC1			1,163 (59)	1,130 (60)	1,129 (65)	144 (59)
SEC2				1,166 (66)	1,154 (61)	144 (71)
SEC3					1,167 (64)	144 (69)
SED						1,109 (81)
SEE						
SEG						
SPE A						
SPE B						
SPE C						
TSST-1						
TSST-O						

Optimized homology scores were calculated using the algorithm of Pearson and Lipman with the FASTA program (ktup = 1) [67]. A Monte Carlo analysis was done using the RDF2 program [67]. In parentheses is the maximum optimized homology score between the test sequence and 100 randomly permutated versions of the query sequence. A test and query sequence have significant homology if the observed score between the test and query sequence is much greater than the maximum score between the test and shuffled

group (table 3). There is little if any significant homology between SPE B or the TSS toxins with any of the other toxins (table 3).

The same general relationships determined by analysis of protein sequences are observed by comparison of nucleic acid sequences of the structural genes (FASTA, ktup = 6; RDF2, 40 random shuffles; data not shown). For those genes that have significant nucleotide sequence identity, the percentage of identity was determined using the GAP program (Genetics Computer Group, Madison, Wisc., USA) set to default parameters (fig. 1) [68]. There is significant nucleotide sequence homology among all the enterotoxins with the exception that *seg⁺* does not have significant homology to either *sed⁺*, *see⁺* or *sezA⁺*. *speA* is related to all the enterotoxins with the greatest homology occurring with *seb⁺*, *sec⁺* and *seg⁺*. *speC* has significant nucleotide sequence homology with *speA* but not with the enterotoxin genes.

SEE	SEG	SPE A	SPE B	SPE C	TSST-1	TSST-O
1,018 (57)	105 (63)	100 (67)	32 (75)	79 (72)	62 (59)	54 (62)
123 (65)	207 (70)	263 (69)	25 (52)	98 (70)	53 (49)	55 (57)
147 (61)	222 (72)	214 (60)	38 (59)	63 (63)	50 (58)	55 (56)
147 (71)	222 (58)	214 (96)	38 (56)	63 (62)	50 (58)	55 (53)
147 (67)	226 (60)	214 (61)	38 (64)	63 (57)	50 (54)	55 (52)
724 (60)	78 (67)	113 (60)	35 (56)	89 (56)	52 (60)	52 (53)
1,153 (57)	99 (55)	104 (55)	31 (52)	85 (58)	63 (64)	65 (52)
	1,179 (59)	263 (58)	25 (57)	81 (54)	56 (55)	64 (54)
		1,110 (58)	33 (61)	129 (62)	57 (68)	65 (60)
			1,224 (53)	29 (57)	39 (50)	39 (68)
				1,030 (66)	35 (59)	42 (68)
					934 (52)	906 (61)
						943 (64)

sequences. A test and query sequence have no significant homology if the observed score is less than the maximum score for the shuffled sequences. A test and query sequence, at best, have limited homology if the observed score is a few points higher than that of the shuffled sequences; for determination of relationships among these toxins, such pairs were considered not to have significant homology. References for each sequence are given in table 1.

Taking the above comparisons into consideration, most of the pyrogenic toxins can be loosely divided into 2 categories. One category has *sea+*, *see+*, *sezA+*, and *sed+*. The other category has *seb+*, *sec+*, *seg+* and *speA*, which are more closely related to one another than to the other toxins. Upon closer inspection, these genes can be classified into 6 groupings as illustrated in figure 1.

Alleles of sec+

Serological examination of SEC produced by various strains led to the recognition of 3 subtypes [9]. The *sec+* genes for 4 strains have been sequenced [42–44, 57]. *S. aureus* strains MN Don, FRI361 and FRI913 produce SEC1, SEC2 and SEC3, respectively [42, 44, 57]. FRI1230 was initially identified as an SEC3 producer based mainly on serological reactions [62]; however, we consider it an SEC2 producer because the nucleotide sequence of *sec+* of strain

Table 3. Percent amino acid sequence identity between mature forms of the toxins given in table 1

Test sequence	Query sequence					
	SEA	SEB	SEC1	SEC2	SEC3	SED
SEA	100 (233)	34 (241)	29 (239)	31 (240)	32 (231)	53 (225)
SEB		100 (239)	66 (241)	66 (241)	66 (241)	36 (238)
SEC1			100 (239)	97 (239)	96 (239)	32 (232)
SEC2				100 (239)	98 (239)	32 (232)
SEC3					100 (239)	32 (232)
SED						100 (228)
SEE						
SEG						
SPE A						
SPE B						
SPE C						
TSST-1						
TSST-O						

Percent identities were obtained using the program FASTA (ktup = 1) [67]. The length of the amino acid overlap region with the reported percent identity is given in parentheses.

FRI1230 (designated $sec^+_{FRI1230}$) is identical to that of sec^+_{FRI361} and different from those of $sec^+_{MN\,Don}$ and sec^+_{FRI913} [42–44, 57]. Purified SEC2 from FRI361 and FRI1230 cultures is not identical. Mature SEC2 purified from FRI361 cultures is 239 residues in length with an N-terminal residue of glutamic acid, whereas the mature form of SEC2 of FRI1230 lacks this N-terminal glutamic acid residue and is only 238 residues in length [42, 43]. It is not clear if the differences between the mature forms of SEC2 are due to strain differences or occur during protein purification. Among the various types of sec^+, there is 97–98% nucleotide sequence identity and 96–98% amino acid sequence identity [42–44, 57] (table 3). The degree of nucleotide and amino acid sequence identity among the pairs correlates with immunological related-ness; SEC2 and SEC3 are the most closely related [44].

Alleles of Toxic Shock Syndrome Toxin Genes

Examination of S. aureus strains associated with mastic disease in animals led to identification of variant forms of toxin that have different isoelectric points from TSST-1 but were indistinguishable in Ouchterlony

SEE	SEG	SPE A	SPE B	SPE C	TSST-1	TSST-O
81 (230)	28 (235)	34 (219)	–	25 (176)	24 (106)	24 (106)
32 (237)	46 (239)	50 (236)	–	23 (197)	20 (152)	19 (152)
29 (230)	42 (238)	47 (235)	23 (92)	26 (125)	30 (60)	32 (60)
31 (231)	42 (238)	46 (235)	23 (92)	26 (125)	30 (60)	32 (60)
31 (231)	42 (238)	47 (235)	23 (92)	26 (125)	30 (60)	32 (60)
56 (225)	27 (235)	37 (223)	27 (37)	31 (159)	25 (111)	24 (111)
100 (230)	28 (233)	35 (219)	–	29 (159)	25 (65)	25 (65)
	100 (233)	44 (238)	–	27 (125)	27 (74)	28 (74)
		100 (221)	–	31 (163)	22 (107)	23 (107)
			100 (253)	–	26 (31)	26 (31)
				100 (208)	21 (97)	22 (89)
					100 (194)	96 (194)
						100 (194)

– = Optimized homology score of best alignment was equal to or less than the cutoff value of 33.

and Western blot assays [24, 25]. TSST-1, which is produced by *S. aureus* strains associated with humans, has a pI of 7.0 [24]. In contrast, ovine-mastic and caprine-mastic isolates produce a variant of TSST-1 with a pI of 8.6 (designated TSST-O) [24]. Another variant form of TSST-1 that has a pI of 7.2 was produced by a bovine isolate [25]. The genes for TSST-1 and TSST-O differ by 21 nucleotides (97% identity) and 9 derived amino acid residues (92% identity) [Kreiswirth and Schlievert, unpubl. data].

Alleles of speA

While examining *S. pyogenes* strains from patients with severe streptococcal disease, 3 new *speA* alleles were identified by Nelson et al. [16]. Five of the isolates had *speA* alleles identical to that reported by Weeks and Ferretti [56]; this allele is designated *speA1* [16]. The allele designated *speA2* differs from *speA1* by a single transition (G → A) at nucleotide 328, resulting in a codon change (glycine → serine) [16]. *speA3* is identical to *speA1*, except for a change at nucleotide 316 which also represents a codon change (G → A; valine → isoleucine) [16]. *speA4* has 67 nucleo-

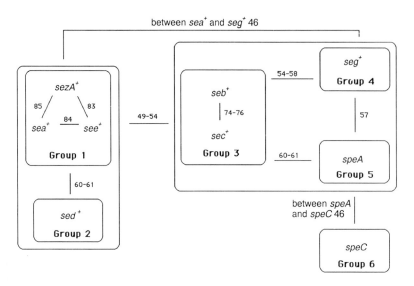

Fig. 1. Sketch illustrating the relationships among the enterotoxins, and SPE A and SPE C genes. The lines between boxes indicate significant nucleotide sequence identity (determined with the computer program FASTA [67]) between all members in both boxes, with two exceptions. *seg⁺* has significant nucleotide sequence identity with *sea⁺* and not with *see⁺*, *sezA⁺* or *sed⁺*. The other exception is *speC* which has significant nucleotide sequence identity only with *speA*. Next to the lines, the percentage of nucleotide sequence identity over the entire structural genes is given (determined with the computer program GAP [68]).

tides (91% nucleotide sequence identity) and 26 inferred amino acid residues (89% amino acid sequence similarity) that are different from *speA1* [16]. Unlike the protein products of the other *speA* alleles, the *speA4* protein product was not detected in an Ouchterlony assay using antibodies produced against SPE A1 [16]. It is not clear if the lack of detection is due to low expression of *speA4* or if the product is antigenically distinct from the other SPE As.

Elements Containing Pyrogenic Toxin Genes and Their Locations

Many of the pyrogenic toxin genes are part of elements, that is a specific toxin gene is flanked by DNA that is not present in most strains which lack that particular gene. The best characterized pyrogenic toxin gene encoding elements are: the streptococcal phages that carry *speA* or *speC*; the family of staphylococcal phages that carry *sea⁺*, *see⁺* and *sezA⁺*, and the *sed⁺*-containing plasmid. The nature of the chromosomal elements containing either *tst* or *seb⁺* is not as well defined.

Elements Containing Staphylococcal Enterotoxin Genes

The first evidence implicating a chromosomal location for *sea⁺* was that SEA production is not associated with detectable plasmid DNA [69]. Conventional transformation techniques demonstrated that the Sea⁺ phenotype is linked to the *purine (pur)* and *isoleucine-valine (ilv)* chromosomal linkage group in 24 of the 29 *S. aureus* strains examined [70, 71]. The fact that for 5 strains Sea⁺ does not co-transfer with the Pur⁺ and Ilv⁺ markers or with any of the other previously mapped linkage groups suggests that *sea⁺* or a gene required for its expression might be on a mobile genetic element [71]. Southern blot analysis with a *sea⁺*-specific probe was used to obtain physical evidence that the structural gene, *sea⁺*, is located within the *pur-ilv* region only for those strains where Sea⁺ co-transfers with the Pur⁺-Ilv⁺ markers [37].

Comparison of Southern blotting patterns between Sea⁺ and Sea⁻ strains is consistent with *sea⁺* being part of a discreet element 8–12 kilobase pairs (kbp) in size [37]. In an effort to establish the identity of this element, early reports about SEA production being phage-associated were re-examined [72, 73]. A phage isolated from *S. aureus* PS42-D (designated phage PS42-D) has DNA that hybridizes to an *sea⁺*-derived probe; this demonstrates that the structural and not a regulator gene for SEA is contained on a phage [74]. Phage PS42-D contains a linear DNA molecule 42 kbp in size with cohesive ends and the phage attachment site *(attP)* located near the middle of the molecule [74]. These features as well as comparison of blotting patterns between DNA from free phage and lysogens suggest that, upon infection, the phage DNA integrates into the chromosome by circularization and reciprocal crossover (Campbell model) [75]. Based on the genetic and physical data, the *sea⁺*-containing phages have a preferred chromosomal integration site within the *pur-ilv* region [70, 71, 74].

sea⁺ is associated with a family of similar but not identical phages. This is based on the finding that *sea⁺*-containing staphylococcal strains have substantial DNA homology with phage PS42-D; however, these phage-related sequences differ by restriction fragment length polymorphisms (RFLPs) [74]. Southern blot analysis also revealed the existence of phages having homology to the *sea⁺*-containing phages but lacking an enterotoxin gene [74]. Viable *sea⁺*-containing phages have been isolated from 6 different wild-type staphylococcal strains, but attempts to isolate *sea⁺*-containing phages from all Sea⁺ strains have not been successful [74, 76]. The inability to detect *sea*⁺-containing phages in all Sea⁺ strains is thought to be due to the presence of defective phages in some strains or phages with altered host ranges [72–74].

An association between *sea*⁺, the *staphylokinase* gene *(sak)* and the lack of β-hemolysin production (Hlb⁻) was observed for 7 of 9 methicillin-resistant *S. aureus* strains isolated from septicemic cases [77]. Because *sea*⁺ and *sak* had each been shown to be phage-associated and the *sak*-containing phage was known to disrupt *hlb* upon integration, the possibility that *sea*⁺ and *sak* were contained by the same phage was investigated [74, 78–80]. Phages from 2 of the *S. aureus* septicemic isolates and from strain PS42-D are similar to a previously characterized phage, phage φ13, in that they belong to serological group F, contain *sak* and integrate into *hlb* [76]. However, they are unlike phage φ13 in that they contain *sea*⁺ [76]. *sak* is located between *sea*⁺ and *attP* with *sea*⁺ located about 6 kbp from *attP* [76]. Detailed cloning and sequence analysis revealed that the chromosomal insertion site *(attB)* for phage φ13 is within *hlb* and has a 14-bp core sequence in common with *attP* of phage φ13 [81]. The ends of the integrated linear prophage are identical to the *attB* and *attP* sequences, indicating that integration of phage φ13 into *hlb* is site- and orientation-specific [81]. Phage PS42-D inactivates *hlb* in an identical manner [D.C. Coleman, pers. communication].

see⁺ is part of a phage that is related to the family of *sea*⁺-containing phages [40]. Two See⁺Sea⁻ strains were examined, and it was found that the UV-induced lysates from these strains are enriched for DNA which hybridizes to *see*⁺ as well as DNA which hybridizes to phage PS42-D [40]. Cloning and sequence analysis of DNA purified from the UV-induced lysate of strain FRI918 provided conclusive evidence that *see*⁺ is associated with this UV-inducible DNA. Apparently, the *see*⁺-containing DNA in the lysates are defective phages, because these UV-induced lysates have no plaque-forming units [40]. The possibility has not been totally discounted that the UV-induced lysates contain 2 different phages, one containing *see*⁺ and another phage that hybridizes to *sea*⁺-containing phage DNA [40].

Another member of the family of *sea*-like containing phages is the *sezA*⁺-containing phage, phage FRI1161 [18]. *sezA*⁺ was discovered in FRI1161 (Sea⁻Sed⁺), a strain that has 2 different DNA fragments that hybridize to *sea*⁺ probes under conditions of reduced stringency [18]. FRI1169 has both a plasmid containing *sed*⁺ and a viable phage that includes *sezA*⁺ [18]. Phage FRI1161 has substantial overall DNA homology with *sea*⁺-containing phages [18]. *sezA*⁺ is not unique to FRI1106; strain FRI377 also has *sezA*⁺ [Borst and Betley, unpubl. results].

sed^+ is contained on a 27.6-kbp plasmid [41]. At least 21 Sed$^+$ strains have been examined and they all have sed^+-containing plasmids with identical restriction patterns [4, 18, 41]. pIB485 of strain 485 is the prototype [41]. This plasmid is related to *S. aureus* penicillinase plasmids; in addition to the penicillin resistance (Pc^r) genes, these plasmids encode cadmium resistance [41].

Early investigations generated a great deal of controversy regarding the location of seb^+. Co-elimination and co-transfer of Pc^r, methicillin resistance (Mc^r) and Seb$^+$ led to the suggestion that seb^+ was plasmid-associated in strains such as DU4916 [82, 83]. Similar linkage among Pc^r, Mc^r, tetracycline resistance (Tc^r) and Seb$^+$ were observed by others [84, 85]. Next, it was shown that the Mc^r marker is chromosomal [86, 87] and not physically linked to Seb$^+$ [84, 85, 88]. Plasmid profiles of DU4916 and its derivatives revealed that Pc^r and Tc^r markers are contained on separate plasmids [85]. The observation that there was 100% correlation between Seb$^+$ and a cryptic 7.5×10^5 dalton plasmid (designated pSN2) led to the speculation that pSN2 was required for SEB production [85]. Subsequently, it has been proven that pSN2 does not contain seb^+ or a function required for seb^+ expression. Evidence for this includes the following: (1) most Mc^r Seb$^+$ strains lack pSN2 [84]; (2) SEB is not produced by either mini cells or a transcription-translation system that is programmed with pSN2 [89]; (3) elimination of pSN2 does not affect SEB production by DU4916 [90], and (4) nucleotide sequence analysis showed that seb^+ was not contained within pSN2 [90].

There is one report of a clinical isolate, strain 6344, containing a 56.2-kbp plasmid, pZA10, which has a gene for Pc^r and also apparently contains seb^+ and sec^+ [91]. The evidence that pZA10 contains seb^+ and sec^+ is that SEB and/or SEC are produced by about 10% of the Pc^r transformants that are produced by transformation of RN450 with pZA10 [91]. However, there is no direct physical evidence that pZA10 contains either enterotoxin gene. Several Seb$^+$ strains have no detectable plasmid DNA associated with seb^+ expression; therefore, seb^+ is assumed to have a chromosomal location in these strains [92].

Physical evidence that seb^+ is contained within a discreet DNA region which is not present in strains which are nonproducers of SEB comes from Southern hybridization experiments [93]. The downstream junction between the seb^+-containing element and the non-seb^+ region has been localized to a region which is 2.3–3.1 kbp downstream from the 5′ end of seb^+ [93]. The exact upstream junction of the seb^+ element has not been localized to the same extent; however, it has been shown that the junction does not lie within 2.9 kbp upstream of the 5′ end of seb^+ [93]. Therefore, these experiments indicate that

seb⁺ is located on a DNA element which is at least 5.2 kbp in size [93]. The
investigators suggested that the *seb*⁺ element may be at least 26.8 kbp in size
[93]. This was based on the observation that an upstream probe corresponding
to the DNA region between 2.9 and 1.8 kbp upstream of *seb*⁺ hybridizes to the
same 24-kbp fragment in all 8 Seb⁺ strains which they examined, whereas they
expected some restriction fragment heterogeneity between the chromosomes of
various Seb⁺ strains [93]. Whether the *seb*⁺ element is an integrated plasmid or
phage is still unresolved [93]; however, no linkage between Seb⁺ and UV-
inducible prophage has been observed for the strains examined [94].

Among the enterotoxin genes, the location of *sec*⁺ is the least well
characterized. As mentioned earlier, pZA10 apparently contains *sec*⁺ [91]. In
all other strains examined, *sec*⁺ was not associated with detectable plasmids
[39, 95] and is, therefore, assumed to be chromosomally located.

Location of Toxic Shock Syndrome Toxin Genes

S. aureus isolates from TSS cases have lysogenic phage normally lacking in
strains not associated with TSS, suggesting that one or more lysogenic phages
might be responsible for the pathogenesis of TSS [96]. No specific phenotype
has been linked to the lysogenic phage associated with the TSS isolates;
therefore, the significance of the observation is difficult to assess.

There is no physical evidence indicating that the TSS toxin gene *(tst)* is
associated with plasmid or phage DNA. Most Tst⁺ strains contain no detectable plasmid DNA [97]. Indicator strains lysogenized with phage isolated from
Tst⁺ strains do not simultaneously acquire the ability to produce TSST-1 [45].

Data from Southern blotting are consistent with *tst* being part of an element
of 4–7 kbp in size that is absent or rearranged in Tst⁻ strains [98]. One boundary
between the element and adjacent chromosomal DNA is about 1.2 kbp from *tst*
and the other boundary has been localized within a 4-kbp EcoRI fragment [98].
Among 13 Tst⁺ isolates, 6 different sizes were observed for the ClaI junction-
containing fragment [98], suggesting that the *tst* element is either flanked by
DNA that has a high degree of restriction site polymorphisms or that the
element has more than one location [98]. The second suggestion is supported by
genetic mapping and phenotypic analysis of Tst⁺ strains [99, 100].

Examination of the *thrB-trp-thy-tyrB* chromosomal linkage group for *tst*
was given impetus by the observation that, unlike Tst⁻ strains, most Tst⁺
S. aureus strains are tryptophan auxotrophs [99]. *tst* of strains S411 (Trp⁻Tst⁺)
and FRI1169 (Trp⁺Tst⁺) were mapped using conventional transformation
procedures [100]. As predicted, *tst* of S411 is located between *thrB* and *thy*,
consistent with the reason for tryptophan auxotrophy being due to the

insertional inactivation of *trp* by the *tst* element [100]. *tst* of FRI1169 is adjacent to *tyrB* on the side opposite of *thy* [100]. Among Tst⁺Trp⁻ strains, there are differences in the ability of tryptophan analogues and intermediaries to satisfy the tryptophan requirement. This suggests that there are differences in the defects in the tryptophan biosynthetic pathway. Such differences may be due to a variation in the insertion site of the *tst* element [100]. Given that *tst* was associated with an element that apparently had more than one chromosomal location and because there is no evidence for this element being a phage, plasmid or transposon, the *tst* element has provisionally been designated as heterologous insertion 555 (i.e. ΩHi555) [100].

Location of Streptococcal Pyrogenic Toxin Genes

In the 1920s, it was reported that a filterable agent from scarlatinal strains converts nonscarlatinal strains into exotoxin producers, suggesting that streptococcal exotoxin production is a phage-associated property [101]. These findings were confirmed by the demonstration that non-exotoxin-producing streptococci acquire the ability to produce exotoxin when lysogenized by temperate phage from strains T12 or 3GL16 [102, 103]. Since all lysogens acquire the ability to produce SPE A and none of the lysogens are Spea⁻, it was most likely that exotoxin production was acquired by lysogenic conversion rather than transduction [104]. Although not all streptococcal phages can transfer Spea⁺ [105], Spea⁺ association with phage is a general phenomenon; in addition to phages T12 and 3GL16, 9 other phages have been shown to be capable of converting Spea⁻ strains to SPE A producers [106]. The Spea⁺-associated phages are all similar in having isometric polyhedral heads, but they differ with respect to host range and there are at least 2 different serotypes [106].

Proof that *speA* is carried by a phage came from analysis of recombinant clones. Two research groups cloned *speA* using phage T12 DNA as the donor and an *E. coli* host/vector system [46, 47]. The recombinant *E. coli* produces SPE A which is capable of inducing T cell proliferation, is biologically active in erythematous skin and in pyrogenicity assays using rabbits, and is serologically identical to SPE A produced by *S. pyogenes* in an Ouchterlony assay [46, 47].

Restriction enzyme analysis revealed that DNA from phage T12 preparations is 35 kbp in length, cyclically permutated and terminally redundant, suggesting that phage T12 may be packaged by a 'headful' cutting mechanism similar to *E. coli* phage T4 and *Salmonella* phage P22 [107]. *speA* is located within 850 bp of *attP* [108].

There is a family of phages that share DNA homology but differ by restriction fragment patterns and the presence of *speA* [108]. Southern blotting analysis using phage T12 DNA as a probe revealed that, out of the 5 Spea⁺ strains examined, all contain phage-T12-related sequences in addition to *speA* [108]. Three Spea⁺ strains which do not contain detectable viable phage have deletions compared with phage T12 [108]. None of 9 Spea⁻ strains that were examined have DNA that hybridizes to *speA*-specific probes, but 5 of these strains have DNA that hybridizes to phage T12, indicating that these strains have a T12-like phage that lacks *speA* [108].

It has been reported that an Spec⁻ strain can become Spec⁺ by lysogenic conversion [109]. *speC* was shown to be part of phage CS112 and to be located near *attP* by Southern blot analysis using *speC*-specific probes obtained from a cloned *speC* [49].

There are conflicting reports about Speb⁺ being phage-associated. Lysogenization of an indicator strain with phage CS112, obtained from an Speb⁺ and Spec⁺ strain, always coincides with SPE C production but none of the lysogens acquire the ability to produce SPE B [104]. When indicator strain T18P (Spea⁻, Speb⁻) is lysogenized with 2 different phages, it acquires the ability to produce detectable amounts of SPE A and SPE B [106]. In a recent report, it has been observed that all 512 group A streptococcal strains examined hybridize to an oligonucleotide probe that corresponds to a portion of *speB*, suggesting that all streptococcal strains may have a form of *speB* [110]. Much work remains to be done concerning the location of *speB*.

Distribution of Streptococcal Pyrogenic Exotoxin Genes and Toxic Shock Syndrome Toxin Genes among Natural Bacterial Populations

Multilocus enzyme electrophoresis is being used to analyze genetic variation among natural populations of bacteria and to construct a genetic framework for use in interpreting variation in specific phenotypic characteristics among the population. This technique has been applied to Tst⁺ *S. aureus* strains and to *S. pyogenes* isolates producing streptococcal pyrogenic exotoxin [111–113].

Among 2,077 *S. aureus* isolates, 252 distinct electrophoretic types (ETs) representing distinct multilocus clonal genotypes have been identified [112]. Based on the observation that 73% of the TSST-1-producing isolates, including about 90% of isolates from patients with menstrual TSS, are in a few closely related ETs, it has been suggested that descents of a single clone of *S. aureus* cause the majority of TSS cases [111, 112].

Thirty-three ETs, representing distinct multilocus clonal genotypes, have been identified among 108 *S. pyogenes* isolates from patients with invasive streptococcal disease [113]. Almost half of these isolates are in either ET1 or ET2, which are 2 closely related groups [113]. Also, ET1 and ET2 contain more than two thirds of the isolates from toxic-shock-like syndrome [113]. This report also contributes to the growing volume of evidence that a high percentage of streptococcal isolates from toxic-shock-like syndrome patients are Spea$^+$ [32–36, 113]. The authors argue that, because evolutionary convergence to the same multilocus genotype is highly unlikely, isolates in a given ET are descended from a common ancestral cell [113]. Thus, the majority of cases of streptococcus-induced toxic-shock-like syndrome are the result of descendants of a single cell [113].

Production of SPE A, SPE B and SPE C among the *S. pyogenes* ETs was examined [113]. Most of the ETs contain some strains that are Speb$^+$, consistent with possession of *speB* being a relatively primitive condition in *S. pyogenes* which has been retained during the course of evolution [113]. This suggestion is supported by data from another research group that found *speB* in almost all 500 streptococcal strains examined [110]. *speC* is also distributed among all the ETs [113]. In contrast, *speA* is confined to phylogenetically related lineages A–C which contain ETs 1–22 [113].

Staphylococcal Toxin Gene Expression

A comprehensive understanding of toxin gene expression includes: (1) recognition of environmental and other conditions (e.g. growth phase) that affect gene expression; (2) identification of elements that are either essential for or that modulate expression; (3) characterization of promoter and, if appropriate, operator regions, and (4) identification of significant differences among wild-type strains. Such basic information is central to understanding bacterial pathogenesis, especially for an organism like *S. aureus* which may be either a benign member of the normal flora or a deadly pathogen.

Expression among the pyrogenic toxin genes is not identical. For instance, early physiological studies have revealed differences in toxin production by staphylococci. For example, SEA and SED are produced primarily during exponential growth [114, 115]. In contrast, the extracellular accumulation of SEB, SEC and TSST-1 is the greatest during the transition into the stationary phase [10, 114, 116]. Among the major serological types of

```
                                18                          +1    +87
  sea⁺  cttttattaTAGACAaatataaaaagtgtatagTAATATatgtatg....ATG

                               16                           +1    +42
  seb⁺  tttttttaaTTGAATatttaagattataacaTATATTtaaagtgtat....ATG

                              14                            +1    +265
  sed⁺  cgatgcgtccggcgTAGAGGatcaaatatattgaTATAATgaaagtg....ATG


                          -35          17±1          -10
  E. coli               ---TTGACA----------------TATAAT---
  consensus
```

Fig. 2. Staphylococcal enterotoxin promoter regions. The site of transcriptional initiation (indicated by +1) was determined by S1 nuclear or primer extension analysis [41, 121; Borst and Betley, unpubl. results] and by comparison to the E. coli consensus promoter sequence. The –10 and –35 promoter regions are in bold print. The distance in nucleotides between the transcriptional and translational initiation sites is indicated for the enterotoxin genes. The distance between the –10 and –35 regions of the promoters is indicated in italics.

enterotoxins, there are differences in toxin yields. No wild-type Sea⁺, Sed⁺ or See⁺ strains have been identified that produce more than 10 µg of enterotoxin per milliliter of culture supernatant [117]. It is not unusual, however, for wild-type Seb⁺ and Sec⁺ strains to produce more than 100 µg/ml [117]. Genetic and molecular techniques are being used to elucidate the underlying mechanisms affecting staphylococcal toxin gene expression. (This article does not include studies that only examined the effects of cultural conditions on accumulated extracellular toxin concentrations [for reviews, see ref. 10 and 117].)

Staphylococcal Enterotoxin Gene Promoters

The promoters for several of the staphylococcal enterotoxin genes have been defined by localization of the transcriptional start sites for the genes and by comparison to the E. coli consensus promoter sequence (fig. 2).

For sea⁺, the site of transcriptional initiation was determined by primer extension analysis and was located 87 ± 1 nucleotides upstream of the translational initiation site [Borst and Betley, unpubl. results]. A region closely resembling an E. coli consensus promoter sequence [118] was located 5 bp upstream of the putative transcriptional initiation site. Deletion analysis of upstream DNA showed that the region of DNA 80 nucleotides upstream from the transcriptional initiation site was sufficient for transcription and expression of sea^+ [Borst and Betley, unpubl. results].

S1 nuclease protection experiments localized the 5' end of the seb^+ mRNA to 42 ± 1 nucleotides upstream from the translational initiation site for seb^+ [120]. Based on a comparison to E. coli consensus promoters [119], the region of DNA 8 nucleotides upstream of the transcriptional initiation site was identified as the putative promoter. Deletion analysis of upstream regions of seb^+ indicated that DNA located 59 nucleotides upstream of the transcriptional initiation site was insufficient for transcription and expression of seb^+; however, a 93-bp region of DNA was sufficient for both transcription and expression of seb^+ [121].

For sed^+, S1 nuclease protection experiments localized the transcriptional start site at 265 ± 1 nucleotides upstream from the translational initiation site for sed^+ [41]. Six nucleotides upstream of the transcriptional initiation site is a sequence resembling an E. coli consensus promoter sequence, although the spacing of 14 nucleotides between the −35 and −10 regions is less than the consensus spacing of 17 ± 1 nucleotides [118].

Overview of the Accessory Gene Regulator

Gene expression of many staphylococcal exoproteins and several cell-surface-associated proteins is regulated by a locus designated the *accessory gene regulator (agr)* [122–125]. Maximal expression of several of the pyrogenic toxin genes requires a functional *agr*. *agr* may also play a role in strain variation in toxin production and in the decreased expression of toxin genes in glucose-containing medium or at alkaline pH. Therefore, a brief description of *agr* is appropriate [for *agr* reviews, see ref. 126–129].

agr is a polycistronic locus containing at least 2 transcripts, RNAII (3.5 kb in size) and RNAIII (0.5 kb in size) [126–129]. RNAII and RNAIII are transcribed in divergent directions and their transcriptional start sites have been reported, by different research groups, to be 184 and 187 bp apart [126, 129]. DNA corresponding to RNAII contains 4 open reading frames required for *agr* function, including *agrA* and *agrB* [129]. Amino acid sequence comparison revealed that the predicted products of *agrA* and *agrB* have significant homology to signal-transducing response regulators and histidine protein kinases, respectively [129]. The DNA corresponding to RNAIII contains the structural gene for δ-hemolysin *(hld)* [130]. RNAII expression appears to be autocatalytic as well as required for RNAIII expression [128, 129]. Both RNAII and RNAIII expression are temporally regulated with maximal expression occurring during postexponential growth [128, 129].

agr regulation of target gene expression is not understood. Accumulated data are consistent with there being more than one mechanism by which *agr* regulates target genes. Early observations involving several target genes indicate that *agr*-mediated regulation may either activate or repress target gene expression; postexponential expression of coagulase and protein A genes (*coa* and *spa*, respectively) decreases while that of genes encoding exoproteins, such as α-hemolysin *(hla)*, increases [131]. *agr*-mediated gene activation appears to affect target genes to different degrees; among the toxin genes, TSST-1 is at least 100-fold lower in an Agr⁻ strain compared to an Agr⁺ strain, while SEB, SEC and SED production decreases 4- to 16-fold in Agr⁻ backgrounds as compared to Agr⁺ backgrounds [41, 120, 123, 132]. Janzon and Arvidson [133] demonstrated that the transcript for RNAIII (and not δ-hemolysin) is sufficient for normal *agr* regulation of *hla* and *spa* [133]. However, δ-hemolysin is required for normal expression of serine and metalloproteases [133]. Activation of *hla* expression requires not only RNAIII but also a temporal regulator that is separate from *agr* [134]. In contrast, *spa* expression is not directly affected by this temporal regulator; *spa* production is always turned off when RNAIII is present [134].

agr Regulation of the Expression of Pyrogenic Toxin Genes

The description by Recsei et al. [123] of *agr* as a regulator of numerous proteins included the examination of the *agr* effect on *tst*. Comparison of RN4282 (a TSST-1-producing clinical isolate) to RN4256 (transductant of RN4282 that received *agr*::Tn*551* from ISP546) revealed that TSST-1 production was at least 100-fold lower in the Agr⁻ strain compared to its Agr⁺ parent strain [123]. Addition of a recombinant plasmid containing RNAII and RNAIII genes restores TSST-1 production in RN4256 to nearly wild-type levels [129]. In an Agr⁺ background, maximal levels of steady-state *tst* mRNA are produced during postexponential growth; Agr⁻ strains have no detectable *tst* mRNA [123]. Studies using the *tst* promoter fused to the *blaZ* reporter gene demonstrated that *agr* regulation of *tst* expression is at the level of transcription and that the *agr* product activates expression from the *tst* promoter following exponential growth [127].

Gaskill and Khan [120] demonstrated that *agr* affects *seb* expression at the level of steady-state *seb* mRNA. (The phrase 'steady state' indicates that the procedure used to quantify the mRNA did not distinguish between differences in mRNA concentration due to differential rates of transcription or mRNA decay.) Two pC194 derivatives with differently sized *seb*⁺-containing inserts were each transferred into ISP479 (Agr⁺) and ISP546 (Agr⁻

derivative of ISP479 [120]). The Agr⁺ strains produce 3- to 7-fold more SEB and have about 4-fold more *seb⁺* mRNA compared to their respective (Agr⁻) or native strains [120].

sec⁺ expression is also regulated by *agr* at the level of steady-state *sec⁺* mRNA. Two sets of Agr⁺/Agr⁻ strains were constructed by transferring the *agr*::Tn*551* of ISP546 into 2 SEC-producing strains, FRI400 and MJB466 [132]. The Agr⁺ parent strains produce more SEC than their respective Agr⁻ derivatives: 16- to 32-fold more for one Agr⁺/Agr⁻ pair and at least 4-fold more for the other pair [132]. Each Agr⁺ parent strain has 2- to 3-fold more *sec⁺* mRNA than its respective Agr⁻ derivative [132]. The effect of an Agr⁻ phenotype on *sec⁺* expression can be overcome by complementation with the DNA encoding RNAIII [Regassa and Betley, unpubl. results].

Bayles and Iandolo [41] reported that *sed⁺* expression is affected by *agr*. A derivative of RN4220 (Agr⁺) that had pIB586 (a recombinant plasmid with an insert of 1.3 kbp that contains *sed⁺*) produced 4- to 5-fold more SED compared to an ISP546 (Agr⁻) derivative that contained pIB586 [41]. The authors noted that RN4220 is considered to be a weak Agr⁺ strain, and the *agr* effect on *sed⁺* expression might be greater in a strain with wild-type levels of *agr* expression [41].

agr has no apparent affect on *sea⁺* expression [Tremaine and Betley, unpubl. results]. *agr*::Tn*551* of ISP546 was transduced into 3 different Sea⁺ strains: FRI100, FRI281A and MJB386. Each of the Agr⁺ strains produced the same amount of SEA as its respective Agr⁻ transductant. Experimental controls included demonstrating that the parent strains are Agr⁺, as evidenced by being hemolytic and containing RNAIII, whereas the transductants are nonhemolytic and contain no detectable RNAIII.

Repression of sec⁺ Expression by Alkaline pH Is Mediated through agr

agr is subject to pH-mediated regulation. When a batch culture of *S. aureus* is grown in complex medium under conditions of nonmaintained pH, postexponential *agr* expression remains high until the culture reaches a pH greater than 7.5 [Regassa and Betley, unpubl. results]. The decrease in RNAIII above pH 7.5 is not due to temporal regulation; a second culture maintained at pH 6.5 continues to produce detectable levels of RNAIII 24 h after inoculation [Regassa and Betley, unpubl. results]. When cultures are maintained at pH 6.5, 7.0, 7.5 or 8.0 throughout growth, a similar pattern is observed. Dramatically decreased levels of RNAIII are seen at pH 7.5 and 8.0 as compared to pH 6.5 and 7.0 [Regassa and Betley, unpubl. results]. The decrease in *agr* expression under alkaline pH conditions appears to be passed on to at least one target gene, *sec*⁺. Less accumulated extracellular SEC and less *sec*⁺ mRNA are produced by the cultures grown at pH 7.5 and 8.0 as compared to the cultures grown at pH 6.5 and 7.0 [Regassa and Betley, unpubl. results]. The alkaline pH effect on *sec*⁺ expression is abolished in a strain with an insertionally inactivated *agr* (MJB710, *agr*::Tn*551* [Regassa and Betley, unpubl. results]).

Glucose Effect on Pyrogenic Toxin Gene Expression

In *S. aureus*, several of the studies examining the effect of glucose on gene expression have concentrated on the enterotoxins. Early studies showed that when glucose is added to the growth medium, less extracellular SEA, SEB and SEC are produced as compared to a culture grown in the absence of glucose [135–137]. The same type of glucose effect on TSST-1 production has been reported [138]. The cultures grown in the presence of glucose exhibit a dramatic decrease in pH, presumably due to the end products of glucose metabolism. To determine if the glucose-related decrease in extracellular enterotoxin required the accompanying decline in pH, cultures were examined under conditions of maintained pH. Jarvis et al. [137] demonstrated that decreased production of extracellular SEA, SEB and SEC is still observed when glucose is added to culture medium maintained at a pH of 6.5. Addition of glucose to the growth medium also results in increased osmotic pressure in the glucose-containing culture. However, increasing the osmotic pressure of the growth medium by addition of NaCl does not decrease extracellular SEC concentrations, suggesting that the observed glucose effect is not due to increased osmotic pressure [Regassa and Betley, unpubl. results].

Further characterization of the glucose effect on the enterotoxins has focused on *sec*⁺. The glucose effect on *sec*⁺ expression is evident only during postexponential growth, and regulation occurs at the level of steady-state *sec*⁺ mRNA [132]. An *sec*⁺ strain with an insertionally inactivated *agr* (MJB710, *agr*::Tn*551*) still exhibits glucose-mediated regulation of *sec*⁺ [132], although SEC levels are much lower in the Agr⁻ background compared with an Agr⁺ strain.

Even though *agr* is not required for the glucose effect on *sec*⁺, *agr* can apparently contribute to the observed glucose effect on *sec*⁺ under certain conditions. When glucose is added to the growth medium and the pH of the culture is allowed to decrease, not only is *sec*⁺ expression diminished but *agr* expression also decreases [Regassa and Betley, unpubl. results]. The magnitude of the apparent glucose effect on *sec*⁺ expression is greater under conditions of nonmaintained pH, which leads to decreased *agr* expression [Regassa and Betley, unpubl. results]. Under conditions of maintained pH, *agr* expression is unaffected by the presence of glucose and a less dramatic glucose effect is observed on *sec+* expression [Regassa and Betley, unpubl. results].

The molecular mechanisms involved in the glucose effect in *S. aureus* are not well defined. Glucose affects the expression of a number of genes in addition to the enterotoxins {e.g. the lactose operon *(lac)* [139]}, and there is no evidence to suggest that all of these genes share the same mechanism for glucose-mediated regulation. The glucose effect on the enterotoxins appears to be different than the well-characterized catabolite regression observed in *E. coli* [reviewed in ref. 140]. In *E. coli*, glucose repression can be overcome by the addition of exogenous cAMP to the growth medium [reviewed in ref. 141]. In contrast, addition of cAMP to glucose-containing *S. aureus* cultures does not relieve the glucose effect on extracellular SEA or SEB accumulation [142, 143]. The effect of several glucose analogues on enterotoxin production has also been studied in an attempt to elucidate the necessary intracellular form of glucose; however, the results are difficult to interpret because these analogues often inhibit cell growth [142–144]. A *cis*-acting site of action for a regulatory molecule has been proposed for the *lac* operon which includes both inverted and direct repeats upstream of the transcriptional start site [139].

Variation in Enterotoxin Production among Staphylococcal Strains

Differences in the concentration of extracellular enterotoxin are observed among various strains containing a specific enterotoxin gene. For instance, wild-type strains FRIS6, DU4916 and COL secrete 375, 50 and 12 µg of SEB/ml, respectively [84, 92, 145]. Regulation of *seb+* occurs at the level of transcription or mRNA stability. Southern hybridization analysis indicates that a single copy of *seb+* is present in FRIS6, DU4916 and COL and the sequence of the upstream promoter region for *seb+* is identical in each of these strains [84, 92, 145]. Cloned *seb+s* from these 3 strains were introduced into the same host background and found to produce equal amounts of *seb+* mRNA and extracellular SEB [145]. Compagnone-Post et al. [145] concluded that host factors present in the parental strains are responsible for the variation in SEB production. They suggested that one of the possible host factors responsible for variable expression of *seb+* among strains is *agr*; levels of RNAIII seem to correlate with the levels of *seb+* mRNA, although the magnitude of the correlation is not absolute [145]. There is a similar correlation seen for Sec+ strains; strains that produce high concentrations of extracellular SEC contain high concentrations of *sec*[+] mRNA and RNAIII [Regassa, Couch and Betley, unpubl. results].

Variable *sea*[+] expression has been examined for 2 Sea[+] strains which exhibit the same growth patterns when cultured under identical conditions. Strain differences in *sea*[+] expression occur at the level of steady-state *sea*[+] mRNA [Borst and Betley, unpubl. data]. At least part of the differences in *sea*[+] expression are associated with the *sea*[+]-containing phages [Borst and Betley, unpubl. data]. Two wild-type Sea[+] strains, FRI100 and FRI281A, were examined; FRI100 produces 8-fold more SEA than FRI281A [Borst and Betley, unpubl. data]. UV-inducible phages from FRI100 and FRI281A were used to lysogenize an Sea[-] strain, ISP456. Southern blot analysis confirmed that FRI100, FRI281A and the 2 lysogens each had a single copy of *sea*[+] [Borst and Betley, unpubl. data]. The ISP456 lysogen with the phage from FRI281A produced 2- to 4-fold less SEA than the ISP456 lysogen with the phage from FRI100 [Borst and Betley, unpubl. results]. The magnitude of the difference in SEA production was not as great between the 2 ISP456 lysogens as it was between the wild-type parents, so host background may contribute partially to differences in SEA production between these 2 strains in addition to the phage-associated contribution. The copy number of *sea*[+] was identical and was equal to one for FRI100, FRI281A and for the 2 lysogens [Borst and Betley, unpubl. results].

Concluding Remarks

Staphylococcal enterotoxins, TSS toxins and streptococcal pyrogenic toxins are considered a family of toxins because they share certain immunological properties. Based on nucleotide and derived amino acid sequence identity, the enterotoxin genes *speA* and *speC* are related to one another; these genes have little if any significant sequence identity with either *tst* or *speB*.

speB and its gene product are unique among the toxins covered in this article. *speB* is unlike the other pyrogenic toxins in several ways: (1) *speB* is found in all group A streptococcal strains examined; (2) the mature form of SPE B is not simply released by cleavage of a signal sequence, but it is the result of extensive processing, and (3) some SPE B variants have proteolytic activity. Either the ancestral gene for *speB* and the other pyrogenic toxins diverged a very long time ago, or the common immunological properties *speB* shares with the other pyrogenic toxins are a result of convergent evolution.

The majority of the pyrogenic toxin genes are associated with variable genetic elements. *sea*[+], *see*[+], *sezA*[+], *speA* and *speC* are each phage-associated.

Clearly, phages are a vital force for providing phenotypic diversity among staphylococcal and streptococcal populations. It is not uncommon for staphylococci to contain lysogenic phage [94]. Being part of a phage provides a ready means of horizontal transfer for toxin genes among strains. Not only can acquisition of a phage provide new genes, but it can also inactivate other genes; 2 such examples are the *sea*[+] family of phages and phage L54a which integrate and disrupt β-hemolysin and the lipase genes, respectively [80, 146]. *sed*[+] is part of a plasmid. Transfer of such plasmids would be expected to occur by generalized transduction and at a higher frequency compared to a gene that exists as one copy per cell. *tst* and *seb*[+] are each considered to be part of an element since each is flanked by DNA not present in strains lacking the respective gene. The nature (e.g. transposon, integrated plasmid) of the *tst* element or the *seb*[+] element is not known.

The phages containing *sea*[+], *see*[+] and *sezA*[+] are related based on DNA sequence homology, but they are not identical as evidenced by RFLPs. Some members of this family are defective and others have no enterotoxin-related sequences. Similarly there is a family of streptococcal phages that differ by RFLPs and which may be either viable or defective; some of these phages contain *speA*. Experimental data suggest that the *speA*-containing phage is quite different from the *sea*[+]-containing phages. Specifically, DNA from *speA* phage particles is circularly permutated and terminally redundant, whereas DNA isolated from *sea*[+]-containing phages does not have these properties.

How these toxin genes became associated with the various elements is not known. The model originally proposed to explain acquisition of the diphtheria toxin gene by the corynebacterium phage may be applicable to the pyrogenic toxin genes [147]. Perhaps there was an ancestral chromosomal toxin gene that was acquired by nearby elements through imprecise excision events. Such models have been proposed for phage acquisition of *sea*[+], *speA* and *speC*, and for formation of the *seb*[+]-containing element which may be an integrated phage, plasmid or, as yet, uncharacterized element [46, 49, 74].

A long-term goal of gene expression studies is to understand factors that affect the expression of genes in various natural environments (e.g. virulence factor expression for normal flora and during a septicemic infection). Towards this goal, factors affecting toxin gene expression are being identified. The molecular mechanisms involved in controlling toxin gene expression are only beginning to be explored. For a given gene, there may be multiple mechanisms of control as observed for *hla* expression. In a wild-type strain, *hla* expression requires both *agr* and a temporal regulator [122, 123, 134]. In addition, in the presence of glucose and nonmaintained pH, *hla* expression is reduced by both an *agr*-dependent and an *agr*-independent mechanism.

In considering the enterotoxins and *tst*, there are at least 3 categories of genes based on expression studies. Unlike *sea*[+] (category 1), maximal production of *seb*[+], *sec*[+] and *tst* (category 2) occurs during postexponential growth, and *seb*[+], *sec*[+] and *tst* are regulated by *agr*. *sed*[+] (category 3) is similar to *sea*[+] in that maximal expression occurs prior to postexponential growth, but like the genes of category 2 extracellular SED concentrations are affected by *agr*. These differences with respect to *agr* regulation and time of maximal toxin production relative to growth phase suggest that there are circumstances in the natural environment that may favor production of one type of toxin over another. Perhaps differences among toxin genes and the factors that affect their expression may contribute to certain types of toxins being more frequently associated with a given condition. A few such examples include: (1) *seb*[+] and *sec*[+] are associated with nonmenstrual TSS cases more often than *sea*[+] [5, 26–36]; (2) *sea*[+], and not the other enterotoxin types or TSST-1, is associated frequently with staphylococcal septicemic isolates, and (3) SEA is associated with more food poisoning outbreaks than SEB or SEC [148–151]. With regard to relative concentrations of enterotoxins produced in foods compared with laboratory media, strain FRIS6 (Sea[+], Seb[+]) in laboratory media produces much more SEB than SEA [114]. However, in pasta dough and Manchego-type cheese that had been inoculated with strain FRIS6, SEA and no SEB was detected [152, 153]. These observations support the suggestion that the levels of toxin observed under laboratory conditions may not accurately reflect the amount of toxin produced in some natural environments.

There are a remarkable number of serologically distinct types of pyrogenic toxins. In addition, more than 1 allele has been identified for *sec*[+], *speA* and *tst*. For a given gene, variant alleles may be generated by random mutation or recombination between 2 different genes. For instance, *speA1, speA2* and *speA3* differ from each other by 1 bp, which is probably the result of random mutations [16]. Nucleotide sequence comparison among the *sec*[+] alleles and *seb*[+] supports the hypothesis that variant forms of enterotoxin genes may have formed by recombination between 2 different enterotoxin genes as well as by random mutations. Two examples support this hypothesis. $sec^+_{FRI1230}$ and $sec^+_{MN\ Don}$ differ by 14 bp and most of these differences are confined to a 37-bp segment [43]. The corresponding 37-bp segment of *seb*[+] is identical to that of $sec^+_{MN\ Don}$ [43]. This observation is consistent with the formation of $sec^+_{MN\ Don}$ occurring by recombination between ancestral $sec^+_{FRI1230}$ and *seb*[+]

genes; the 4 other differences between $sec^+_{FRI1230}$ and $sec^+_{MN\,Don}$ may be the result of accumulated random mutations [43]. In another report, comparisons of signal sequences of SEB and SECs lead to the suggestion that recombination has occurred between these genes [44]. Specifically, within the first 27 codons of seb^+ and sec^+_{FRI913} (encodes SEC3), there is only a difference of 1 codon. In contrast, the SEC3 signal sequence differs by 5 codons from the corresponding regions of SEC1 and SEC2, which are identical [44]. Given these observations, new types of pyrogenic toxin genes will continue to evolve.

Acknowledgements

We would like to thank Drs. David Coleman, Barry Kreiswirth, James Musser and Patrick Schlievert for generously providing unpublished results. We are also grateful to Ms. Susan Reis for her excellent technical assistance in preparing this manuscript.

The work was supported in part by Public Health Service grant AI-25574 from the National Institutes of Health and funds from the Hatch Act, administered through the Agricultural Experimental Station at the University of Wisconsin-Madison (M.J.B.). L.B.R. was supported with funds from Cellular and Molecular Biology Training grant 5 T32 GM07215.

References

1 Betley MJ, Miller VL, Mekalanos JJ: Genetics of bacterial enterotoxins. Annu Rev Microbiol 1986;40:577–605.
2 Ferretti JJ, Yu CE, Hynes WL, Weeks CR: Molecular characterization of the group A streptococcal exotoxin type A (erythrogenic toxin) gene and product; in Ferretti JJ, Curtiss R III (eds): Streptococcal Genetics. Washington, American Society for Microbiology, 1987, pp 130–135.
3 Schlievert PM, Johnson LP, Tomai MA, Handley JP: Characterization and genetics of group A streptococcal pyrogenic exotoxins; in Ferretti JJ, Curtiss R III (eds): Streptococcal Genetics. Washington, American Society for Microbiology, 1987, pp 136 142.
4 Iandolo JJ: Genetic analysis of extracellular toxins of *Staphylococcus aureus*. Annu Rev Microbiol 1989;43:375–402.
5 Bohach GA, Fast DJ, Nelson RD, Schlievert PM: Staphylococcal and streptococcal pyrogenic toxins involved in toxic shock syndrome and related illnesses. Microbiology 1990;117:251–272.
6 Betley MJ, Soltis MT, Couch JL: Molecular biological analysis of staphylococcal enterotoxin genes; in Novick RP (ed): Molecular Biology of the Staphylococci. New York, VCH, 1990, pp 327–342.
7 Khan SA, Gaskill ME, Mahmood R, Wong H, Johns MB Jr, Zock JM: Studies on the staphylococcal enterotoxin B gene; in Novick RP (ed): Molecular Biology of the Staphylococci. New York, VCH, 1990, pp 289–299.

8 Schlievert PM, Bohach GA, Hovde CJ, Kreiswirth BN, Novick RP: Molecular studies of toxic shock syndrome-associated staphylococcal and streptococcal toxins; in Novick RP (ed): Molecular Biology of the Staphylococci. New York, VCH, 1990, pp 311–325.

9 Bergdoll MS: Enterotoxins; in Easmon CSF, Adlam C (eds): Staphylococci and Staphylococcal Infections. New York, Academic Press, 1983, pp 559–598.

10 Bergdoll MS, Chesney PJ (eds): Toxic Shock Syndrome. Boca Raton, CRC Press, 1991.

11 Blomster-Hautamaa DA, Schlievert PM: Nonenterotoxic staphylococcal toxins; in Hardegree MC, Tu AT (eds): Bacterial Toxins. New York, Dekker, 1988, pp 297–330.

12 Spero L, Johnson-Winegar A, Schmidt JJ: Enterotoxins of staphylococci; in Hardegree MC, Tu AT (eds): Bacterial Toxins. New York, Dekker, 1988, pp 131–163.

13 Wannamaker LW, Schlievert PM: Exotoxins of group A streptococci; in Hardegree MC, Tu AT (eds): Bacterial Toxins. New York, Dekker, 1988, pp 267–296.

14 Marrack P, Kappler J: The staphylococcal enterotoxins and their relatives. Science 1990;248:705–711.

15 Watson KC: Host-parasite factors in group A streptococcal infections: Pyrogenic and other effects of immunologic distinct exotoxins related to scarlet fever toxin. J Exp Med 1960;111:255–260.

16 Nelson K, Schlievert PM, Selander RK, Musser JM: Characterization and clonal distribution of four alleles of the *speA* gene encoding pyrogenic exotoxin A (scarlet fever toxin) in *Streptococcus pyogenes*. J Exp Med, in press.

17 Schwab JH, Watson DW, Cromartie WJ: Further studies of group A streptococcal factors with lethal and cardiotoxic properties. J Infect Dis 1955;96:14–18.

18 Soltis MT, Mekalanos JJ, Betley MJ: Identification of a phage containing a staphylococcal variant enterotoxin gene *(sezA⁺)*. Infect Immun 1990;58:1614–1619.

19 Betley MJ, Schlievert PM, Bergdoll MS, Bohach GA, Iandolo JJ, Khan SA, Pattee PA, Reiser RR: Staphylococcal gene nomenclature. ASM News 1990;56:182.

20 Parsonnet J, Hickman RK, Eardley DD, Pier GB: Induction of human interleukin 1 by toxic shock syndrome toxin 1. J Infect Dis 1985;151:514–522.

21 Dinarello CA, Cannon JG, Wolff SM, Bernheim HA, Beutler B, Cerami A, Figori IS, Palladino MA, O'Connor JV: Tumor necrosis factor (cachectin) is an endogenous pyrogen and induces production of interleukin 1. J Exp Med 1986;163:1433–1450.

22 Fast DJ, Schlievert PM, Nelson RD: Non-purulent response to toxic shock syndrome toxin-1 producing *Staphylococcus aureus*: Relationship to toxin-stimulated production of tumor necrosis factor. J Immunol 1988;140:949–953.

23 Fast DJ, Schlievert PM, Nelson RD: Toxic shock syndrome-associated staphylococcal and streptococcal pyrogenic toxins are potent inducers of tumor necrosis factor production. Infect Immun 1989;57:291–294.

24 Ho G, Campbell WH, Bergdoll MS, Carlson E: Production of a toxic shock syndrome toxin variant by *Staphylococcus aureus* strains associated with sheep, goats, and cows. J Clin Microbiol 1989;27:1946–1948.

25 Ho G, Campbell WH, Carlson E: Ovine-associated *Staphylococcus aureus* protein with immunochemical similarity to toxic shock syndrome toxin 1. J Clin Microbiol 1989;27:210–212.

26 Milner LS, De Jager J, Thomson PD, Doehring RO: Toxic shock syndrome caused by staphylococcal enterotoxin B. S Afr Med J 1983;63:822–824.

27 Crass BA, Bergdoll MS: Involvement of staphylococcal enterotoxins in nonmenstrual toxic shock syndrome. J Clin Microbiol 1986;23:1138–1139.

28 Crass BA, Bergdoll MS: Toxin involvement in toxic shock syndrome. J Infect Dis 1986;153:918–926.

29 Schlievert PM: Staphylococcal enterotoxin B and toxic-shock syndrome toxin-1 are significantly associated with non-menstrual TSS. Lancet 1986;i:1149–1150.

30 Rizkallah MF, Tolaymat A, Martinez JS, Schlievert PM, Ayoub EM: Toxic shock syndrome caused by a strain of *Staphylococcus aureus* that produces enterotoxin C but not toxic shock syndrome toxin-1. Am J Dis Child 1989;143:848–849.

31 McCollister BD, Kreiswirth BN, Novick RP, Schlievert PM: Production of toxic shock syndrome-like illness in rabbits by *Staphylococcus aureus* 04508: Association with enterotoxin A. Infect Immun 1990;58:2067–2070.

32 Cone LA, Woodard DR, Schlievert PM, Tomory GS: Clinical and bacteriologic observations of a toxic shock-like syndrome due to *Streptococcus pyogenes*. N Engl J Med 1987;317:146–149.

33 Bartter T, Dascal A, Carrol K, Curley FC: 'Toxic strep syndrome': Manifestation of group A streptococcal infections. Arch Intern Med 1988;148:1421–1424.

34 Hribalova V: *Streptococcus pyogenes* and toxic shock syndrome. Ann Intern Med 1988;108:722.

35 Stevens DL, Tanner MH, Winship J, Swarts R, Reis KM, Schlievert PM, Kaplan E: Severe group A streptococcal infections associated with a toxic shock-like syndrome and scarlet fever toxin A. N Engl J Med 1989;321:1–7.

36 Hauser AR, Stevens DL, Kaplan EL, Schlievert PM: Molecular analysis of pyrogenic exotoxins from *Streptococcus pyogenes* isolates associated with toxic shock-like syndrome. J Clin Microbiol, in press.

37 Betley MJ, Lofdahl S, Kreiswirth BN, Bergdoll MS, Novick RP: Staphylococcal enterotoxin A gene is associated with a variable genetic element. Proc Natl Acad Sci USA 1984;81:5179–5183.

38 Ranelli DM, Jones CL, Johns MB, Mussey GJ, Khan SA: Molecular cloning of staphylococcal enterotoxin B gene in *Escherichia coli* and *Staphylococcus aureus*. Proc Natl Acad Sci USA 1985;82:5850–5854.

39 Bohach GA, Schlievert PM: Expression of staphylococcal enterotoxin C_1 in *Escherichia coli*. Infect Immun 1987;55:428–432.

40 Couch JL, Soltis MT, Betley MJ: Cloning and nucleotide sequence of the type E staphylococcal enterotoxin gene. J Bacteriol 1988;170:2954–2960.

41 Bayles KW, Iandolo JJ: Genetic and molecular analyses of the gene encoding staphylococcal enterotoxin D. J Bacteriol 1989;171:4799–4806.

42 Bohach GA, Schlievert PM: Conservation of the biologically active portions of staphylococcal enterotoxins C1 and C2. Infect Immun 1989;57:2249–2252.

43 Couch JL, Betley MJ: Nucleotide sequence of the type C_3 staphylococcal enterotoxin gene suggests that intergenic recombination causes antigenic variation. J Bacteriol 1989;171:4507–4510.

44 Hovde CJ, Hackett SP, Bohach GA: Nucleotide sequence of the staphylococcal enterotoxin C3 gene: Sequence comparison of all three type C staphylococcal enterotoxins. Mol Gen Genet 1990;220:329–333.

45 Kreiswirth BN, Lofdahl S, Betley MJ, O'Reilly M, Schlievert PM, Bergdoll MS, Novick RP: The toxic shock syndrome exotoxin structural gene is not detectably transmitted by a prophage. Nature 1983;305:709–712.

46 Johnson LP, Schlievert PM: Group A streptococcal phage T12 carries the structural gene for pyrogenic exotoxin type A. Mol Gen Genet 1984;194:52–56.

47 Weeks CR, Ferretti JJ: The gene for type A streptococcal exotoxin (erythrogenic toxin) is located in bacteriophage T12. Infect Immun 1984;46:531–536.

48 Bohach GA, Hauser AR, Schlievert PM: Cloning of the gene, speB, for streptococcal pyrogenic exotoxin type B in Escherichia coli. Infect Immun 1988;56:1665–1667.

49 Goshorn SC, Bohach GA, Schlievert PM: Cloning and characterization of the gene, speC, for pyrogenic exotoxin type C from Streptococcus pyogenes. Mol Gen Genet 1988;212:66–70.

50 Kreiswirth BN, Handley JP, Schlievert PM, Novick RP: Cloning and expression of streptococcal pyrogenic exotoxin A and staphylococcal toxic shock syndrome toxin-1 in Bacillus subtilis. Mol Gen Genet 1987;208:84–87.

51 Huang IY, Bergdoll MS: The primary structure of staphylococcal enterotoxin B. III. The cyanogen bromide peptides of reduced and amino-ethylated enterotoxin B. J Biol Chem 1970;245:3518–3528.

52 Schmidt JJ, Spero L: The complete amino acid sequence of staphylococcal enterotoxin C1. J Biol Chem 1983;258:6300–6306.

53 Blomster-Hautamaa DA, Kreiswirth BN, Kornblum JS, Novick RP, Schlievert PM: The nucleotide and partial amino acid sequence of toxic shock syndrome toxin-1. J Biol Chem 1986;261:15783–15786.

54 Johnson LP, L'Italien JJ, Schlievert PM: Streptococcal pyrogenic exotoxin type A (scarlet fever toxin) is related to Staphylococcus aureus enterotoxin B. Mol Gen Genet 1986;203:354–356.

55 Jones CL, Khan SA: Nucleotide sequence of the enterotoxin B gene from Staphylococcus aureus. J Bacteriol 1986;166:29–33.

56 Weeks CR, Ferretti JJ: Nucleotide sequence of the type A streptococcal exotoxin (erythrogenic toxin) gene from Streptococcus pyogenes bacteriophage T12. Infect Immun 1986;52:144–150.

57 Bohach GA, Schlievert PM: Nucleotide sequence of the staphylococcal enterotoxin C1 gene and relatedness to other pyrogenic toxins. Mol Gen Genet 1987;209:15–20.

58 Huang IY, Hughes JL, Bergdoll MS, Schantz EJ: Complete amino acid sequence of staphylococcal enterotoxin A. J Biol Chem 1987;262:7006–7013.

59 Betley MJ, Mekalanos JJ: Nucleotide sequence of the type A staphylococcal enterotoxin gene. J Bacteriol 1988;170:34–41.

60 Goshorn SC, Schlievert PM: Nucleotide sequence of streptococcal pyrogenic exotoxin type C. Infect Immun 1988;56:2518–2520.

61 Hauser AR, Schlievert PM: Nucleotide sequence of the streptococcal pyrogenic exotoxin type B gene and relationship between the toxin and the streptococcal proteinase precursor. J Bacteriol 1990;172:4536–4542.

62 Reiser RF, Robbins RN, Noleto AL, Khoe GP, Bergdoll MS: Identification, purification and some physicochemical properties of staphylococcal enterotoxin C_3. Infect Immun 1984;45:625–630.

63 Gerlach D, Knoll H, Kohler W, Ozegowski JH, Hribalova V: Isolation and characterization of erythrogenic toxins. V. Communication, identity of erythrogenic toxin type B and streptococcal proteinase precursor. Zentralbl Bakteriol Parasitenkd Infektionskr Hyg Abt 1 Orig Reihe A 1983;255:221–233.

64 Konaha K, Elliott SD, Liu TY: Primary structure of zymogen of streptococcal proteinase. J Protein Chem 1982;1:317–334.

65 Tai JY, Kortt AA, Liu TY, Elliott SD: Primary structure of streptococcal proteinase. III. Isolation of cyanogen bromide peptides: Complete covalent structure of the polypeptide chain. J Biol Chem 1976;251:1955–1959.

66 Kokan NP, Bergdoll MS: Detection of low-enterotoxin-producing *Staphylococcus* strains. Appl Environ Microbiol 1987;53:2675–2676.

67 Pearson WR, Lipman DJ: Improved tools for biological sequence comparison. Proc Natl Acad Sci USA 1988;85:2444–2448.

68 Devereux J, Haeberli P, Smithies O: A comprehensive set of sequence analysis programs for the VAX. Nucleic Acids Res 1984;12:387–395.

69 Shafer WM, Iandolo JJ: Staphylococcal enterotoxin A: A chromosomal gene product. Appl Environ Microbiol 1978;36:389–391.

70 Pattee PA, Glatz BA: Identification of a chromosomal determinant of enterotoxin A production in *Staphylococcus aureus.* Appl Environ Microbiol 1980;39:186–193.

71 Mallonee DH, Glatz BA, Pattee PA: Chromosomal mapping of a gene affecting enterotoxin A production in *Staphylococcus aureus.* Appl Environ Microbiol 1982; 43:397–402.

72 Casman EP: Staphylococcal enterotoxin. Ann NY Acad Sci 1965;128:124–131.

73 Jarvis AW, Lawrence RC: Production of extracellular enzymes and enterotoxin A, B and C by *Staphylococcus aureus.* Infect Immun 1971;4:110–115.

74 Betley MJ, Mekalanos JJ: Staphylococcal enterotoxin A is encoded by phage. Science 1985;229:185–187.

75 Campbell AM: Episomes. Adv Genet 1962;11:101–145.

76 Coleman DC, Sullivan DJ, Russell RJ, Arbuthnott JP, Carey BF, Pomeroy HM: *Staphylococcus aureus* bacteriophages mediating the simultaneous lysogenic conversion of β-lysin, staphylokinase and enterotoxin A: Molecular mechanism of triple conversion. J Gen Microbiol 1989;135:1679–1697.

77 Humphreys H, Keane CT, Hone R, Pomeroy H, Russell RJ, Arbuthnott JP, Coleman DC: Enterotoxin production by *Staphylococcus aureus* isolates from cases of septicaemia and healthy carriers. J Med Microbiol 1989;28:163–172.

78 Konodo I, Itch S, Yoshizawa Y: Staphylococcal phages mediating the lysogenic conversion of staphylokinase; in Jeljaszewıcz J (ed): Staphylococci and Staphylococcal Infections. Stuttgart, Fischer, 1981, pp 357–362.

79 Winkler KC, de Waart J, Grootsen C, Zegers BJ, Tellier NE, Vertregt CD: Lysogenic conversion of staphylococci to loss of β-toxin. J Gen Microbiol 1965;39:321–333.

80 Coleman DC, Arbuthnott JP, Pomeroy HM, Birkbeck TH: Cloning and expression in *Escherichia coli* and *Staphylococcus aureus* of the beta-lysin determinant from *Staphylococcus aureus*: Evidence that bacteriophage conversion of beta-lysin activity is caused by insertional inactivation of the beta-lysin determinant. Microb Pathog 1986;1:549–564.

81 Coleman D, Knights J, Russell R, Shanley D, Birkbeck TH, Dougan G, Charles I: Insertional inactivation of the *Staphylococcus aureus* β-toxin by bacteriophage φ13 occurs by site- and orientation-specific integration of the φ13 genome. Mol Microbiol 1991;5:933–939.

82 Dornbusch K, Hallander HO: Transduction of penicillinase production and methicillin resistance-enterotoxin B production in strains of *Staphylococcus aureus.* J Gen Microbiol 1973;76:1–11.

83 Dornbusch K, Hallander OH, Lofquist F: Extrachromosomal control of methicillin resistance and toxin production in *Staphylococcus aureus.* J Bacteriol 1969;98:351–358.

84 Shafer WM, Iandolo JJ: Genetics of staphylococcal enterotoxin B in methicillin-resistant isolates of *Staphylococcus aureus*. Infect Immun 1979;25:902–911.

85 Shalita AI, Hertman I, Sarid S: Isolation and characterization of a plasmid involved with enterotoxin production in *Staphylococcus aureus*. J Bacteriol 1977;129:317–325.

86 Sjostrom JE, Lofdahl S, Philipson L: Transformation reveals a chromosomal locus of the gene(s) for methicillin resistance in *Staphylococcus aureus*. J Bacteriol 1975;123:905–915.

87 Kuhl SA, Pattee PA, Baldwin JN: Chromosomal map location of the methicillin resistance determinant in *Staphylococcus aureus*. J Bacteriol 1978;135:460–465.

88 Lacey RW: *Staphylococcus aureus* strain DU4916, an atypical methicillin-resistant isolate? J Gen Microbiol 1974;84:1–10.

89 Dyer DW, Iandolo JJ: Plasmid-chromosomal transition in genes important in staphylococcal enterotoxin B expression. Infect Immun 1981;33:450–458.

90 Khan SA, Novick RP: Structural analysis of plasmid pSN2 in *Staphylococcus aureus*: No involvement in enterotoxin B production. J Bacteriol 1982;149:642–649.

91 Altboum Z, Hertman I, Sarid S: Penicillinase plasmid-linked genetic determinants for enterotoxins B and C_1 production in *Staphylococcus aureus*. Infect Immun 1985;47:514–521.

92 Shafer WM, Iandolo JJ: Chromosomal locus for staphylococcal enterotoxin B. Infect Immun 1978;20:273–278.

93 Johns MB Jr, Khan SA: Staphylococcal enterotoxin B gene is associated with a discrete genetic element. J Bacteriol 1988;170:4033–4039.

94 Read RB Jr, Pritchard WL: Lysogeny among the enterotoxigenic staphylococci. Can J Microbiol 1963;9:879–889.

95 Betley MJ, Bergdoll MS: Staphylococcal enterotoxin type C genes are not associated with extrachromosomal DNA. Abstr Annu Meet Am Soc Microbiol 1981;1981:49.

96 Schutzer SE, Fischetli VA, Zabriskie JB: Toxic shock syndrome and lysogeny in *Staphylococcus aureus*. Science 1983;220:316–318.

97 Kreiswirth BN, Novick RP, Schlievert PM, Bergdoll M: Genetic studies on staphylococcal strains from patients with toxic shock syndrome. Ann Intern Med 1982;96:974–977.

98 Kreiswirth BN, O'Reilly M, Novick RP: Genetic characterization and cloning of the toxic shock syndrome exotoxin. Surv Synth Path Res 1984;3:73–82.

99 Chu MC, Melish ME, James JF: Tryptophan auxotypy associated with *Staphylococcus aureus* that produce toxic shock syndrome toxin. J Infect Dis 1985;151:1157–1158.

100 Chu MC, Kreiswirth BN, Pattee PA, Novick RP, Melish ME, James JF: Association of toxic shock toxin-1 determinant with a heterologous insertion at multiple loci in the *Staphylococcus aureus* chromosome. Infect Immun 1988;56:2702–2708.

101 Frobisher M, Brown JH: Transmissible toxigenicity of streptococci. Bull Johns Hopkins Hosp 1927;41:167–173.

102 Bingel KF: Neue Untersuchungen zur Scharlachätiologie. Dtsch Med Wochenschr 1949;74:703–706.

103 Zabriskie JB: The role of temperate bacteriophage in the production of erythrogenic toxin by group A streptococci. J Exp Med 1964;199:761–779.

104 Johnson LP, Schlievert PM, Watson DW: Transfer of group A streptococcal pyrogenic exotoxin production to nontoxigenic strains by lysogenic conversion. Infect Immun 1980;28:254–257.

105 McKane L, Ferretti JJ: Phage-host interactions and the production of type A streptococcal exotoxin in group A streptococci. Infect Immun 1981;34:915–919.

106 Nida SK, Ferretti JJ: Phage influence on the synthesis of extracellular toxins in group A streptococci. Infect Immun 1982;36:745–750.

107 Johnson LP, Schlievert PM: A physical map of the group A streptococcal pyrogenic exotoxin bacteriophage T12 genome. Mol Gen Genet 1983;203:354–356.

108 Johnson LP, Tomai MA, Schlievert PM: Bacteriophage involvement in group A streptococcal pyrogenic exotoxin A production. J Bacteriol 1986;166:623–627.

109 Colon-Whitt A, Whitt RS, Cole RM: Production of an erythrogenic toxin (streptococcal pyrogenic exotoxin) by a non-lysogenised group-A streptococcus; in Parker MT (ed): Pathogenic Streptococci. Proc VII Int Symp on Streptococci and Streptococcal Diseases. Chertsey, Reedbooks, 1978, pp 64–65.

110 Yu C, Ferretti JJ: Frequency of the erythrogenic toxin B and C genes *(speB* and *speC)* among clinical isolates of group A streptococci. Infect Immun 1991;59:211–215.

111 Musser JM, Schlievert PM, Chow AW, Ewan P, Kreiswirth BN, Rosdahl VT, Naidu AS, Witte W, Selander RK: A single clone of *Staphylococcus aureus* causes the majority of cases of toxic shock syndrome. Proc Natl Acad Sci USA 1990;87:225–229.

112 Musser JM, Selander RK: Genetic analysis of natural populations of *Staphylococcus aureus*; in Novick RP (ed): Molecular Biology of the Staphylococci. New York, VCH, 1990, pp 59–67.

113 Musser JM, Hauser AR, Kim MH, Schlievert PM, Nelson K, Selander RK: *Streptococcus pyogenes* causing toxic-shock-like syndrome and other invasive diseases: Clonal diversity and pyrogenic exotoxin expression. Proc Natl Acad Sci USA 1991; 88:2668–2672.

114 Bergdoll MS, Czop JK, Gould SS: Enterotoxin synthesis by the staphylococci. Ann NY Acad Sci 1974;236:307–316.

115 Noleto AL, Bergdoll MS: Production of enterotoxin by a *Staphylococcus aureus* strain that produces three identifiable enterotoxins. J Food Prot 1982;45:1096–1097.

116 Otero A, Garcia ML, Garcia MC, Moreno B, Bergdoll MS: Production of staphylococcal enterotoxins C_1 and C_2 and thermonuclease throughout the growth cycle. Appl Environ Microbiol 1990;56:555–559.

117 Bergdoll MS: Staphylococcal intoxications; in Riemann H, Bryan FL (eds): Foodborne Infections and Intoxications. New York, Academic Press, 1979, pp 443–493.

118 Hawley DK, McClure WR: Compilation and analysis of *Escherichia coli* promoter DNA sequences. Nucleic Acids Res 1983;11:2237–2255.

119 Harley CB, Reynolds RP: Analysis of *E. coli* promoter sequences. Nucleic Acids Res 1987;15:2343–2361.

120 Gaskill ME, Kahn SA: Regulation of the enterotoxin B gene in *Staphylococcus aureus.* J Biol Chem 1988;263:6276–6280.

121 Mahmood R, Kahn SA: Role of upstream sequences in the expression of the staphylococcal enterotoxin B gene. J Biol Chem 1990;265:4652–4656.

122 Janzon L, Lofdahl S, Arvidson S: Evidence for a coordinate transcriptional control of alpha-toxin and protein A synthesis in *Staphylococcus aureus.* FEMS Microbiol Lett 1986;33:193–198.

123 Recsei P, Kreiswirth B, O'Reilly M, Schlievert P, Gruss A, Novick RP: Regulation of exoprotein gene expression in *Staphylococcus aureus* by *agr.* Mol Gen Genet 1986;202:58–61.

124 Morfeldt E, Janzon L, Arvidson S, Lofdahl S: Cloning of a chromosomal locus *(exp)* which regulates the expression of several exoprotein genes in *Staphylococcus aureus*. Mol Gen Genet 1988;211:435–440.

125 Peng HL, Novick RP, Kreiswirth B, Kornblum J, Schlievert P: Cloning, characterization, and sequencing of an accessory gene regulator *(agr)* in *Staphylococcus aureus*. J Bacteriol 1988;170:4365–4372.

126 Arvidson S, Janzon L, Lofdahl S, Morfeldt E: The exoprotein regulatory region *exp* of *Staphylococcus aureus*; in Butler LO, Harwood C, Moseley BEB (eds): Genetic Transformation and Expression. Andover, Intercept, 1989, pp 511–518.

127 Novick R, Kornblum J, Kreiswirth B, Projan S, Ross H: *agr*: A complex locus regulating post-exponential phase exoprotein synthesis in *Staphylococcus aureus*; in Butler LO, Harwood C, Moseley BEB (eds): Genetic Transformation and Expression. Andover, Intercept, 1989, pp 495–510.

128 Arvidson S, Janzon L, Lofdahl S: The role of the δ-lysin gene *(hld)* in the agr-dependent regulation of exoprotein synthesis in *Staphylococcus aureus*; in Novick RP (ed): Molecular Biology of the Staphylococci. New York, VCH, 1990, pp 419–431.

129 Kornblum J, Kreiswirth BN, Projan SJ, Ross H, Novick RP: Agr: A polycistronic locus regulating exoprotein synthesis in *Staphylococcus aureus*; in Novick RP (ed): Molecular Biology of the Staphylococci. New York, VCH, 1990, pp 373–402.

130 Janzon L, Lofdahl S, Arvidson S: Identification and nucleotide sequence of the delta-lysin gene, *hld*, adjacent to the accessory gene regulator *(agr)* of *Staphylococcus aureus*. Mol Gen Genet 1989;219:480–485.

131 Bjorkland A, Arvidson S: Mutants of *Staphylococcus aureus* affected the regulation of exoprotein synthesis. FEMS Microbiol Lett 1980;7:203–206.

132 Regassa LB, Couch JL, Betley MJ: Steady-state staphylococcal enterotoxin type C mRNA is affected by a product of the accessory gene regulator *(agr)* and by glucose. Infect Immun 1991;59:955–962.

133 Janzon L, Arvidson S: The role of the δ-lysin gene *(hld)* in the regulation of virulence genes by the accessory gene regulator *(agr)* in *Staphylococcus aureus*. EMBO J 1990;9:1391–1399.

134 Vandenesch F, Kornblum J, Novick RP: A temporal signal, independent of *agr*, is required for *hla* but not *spa* transcription in *Staphylococcus aureus*. J Bacteriol 1991;173:6313–6320.

135 Morse SA, Mah RA, Dubrogosz WJ: Regulation of staphylococcal enterotoxin B. J Bacteriol 1969;98:4–9.

136 Morse SA, Baldwin JN: Regulation of staphylococcal enterotoxin B: Effect of thiamine starvation. Appl Microbiol 1971;22:242–249.

137 Jarvis AW, Lawrence RC, Pritchard GG: Glucose repression of enterotoxins A, B and C and other extracellular proteins in staphylococci in batch and continuous culture. J Gen Microbiol 1975;86:75–87.

138 Schlievert PM, Blomster DA: Production of staphylococcal pyrogenic exotoxin type C: Influence of physical and chemical factors. J Infect Dis 1983;147:236–242.

139 Oskouian B, Stewart GC: Repression and catabolite repression of the lactose operon of *Staphylococcus aureus*. J Bacteriol 1990;172:3804–3812.

140 Saier MH Jr: Protein phosphorylation and allosteric control of inducer exclusion and catabolite repression by the bacterial phosphoenolpyruvate: Sugar phosphotransferase system. Microbiol Rev 1989;53:109–120.

141 Ullmann A, Danchin A: Role of cyclic AMP in bacteria. Adv Cyclic Nucleotide Res 1983;15:1–53.

142 Smith JL, Bencivengo MM, Buchanan RL, Kunsch CA: Enterotoxin A production in *Staphylococcus aureus*: Inhibition by glucose. Arch Microbiol 1986;144:131–136.

143 Iandolo JJ, Shafer WM: Regulation of staphylococcal enterotoxin B. Infect Immun 1977;16:610–616.

144 Smith JL, Bencivengo MM, Buchanan RL, Kunsch CA: Effect of glucose analogs on the synthesis of staphylococcal enterotoxin A. J Food Saf 1987;8:139–146.

145 Compagnone-Post P, Malyankar U, Khan SA: Role of host factors in the regulation of the enterotoxin B gene. J Bacteriol 1991;173:1827–1830.

146 Lee CY, Iandolo JJ: Lysogenic conversion of staphylococcal lipase is caused by insertion of the bacteriophage L54a genome into the lipase structural gene. J Bacteriol 1986;166:385–391.

147 Laird W, Groman N: Prophage map of converting corynebacteriophage beta. J Virol 1976;19:208–219.

148 Wieneke AA: Enterotoxin production by strains of *Staphylococcus aureus* isolated from foods and human beings. J Hyg 1974;73:255–262.

149 Petras P, Maskova L: Detection of staphylococcal enterotoxigenicity. II. Field strains. J Hyg Epidemiol Microbiol Immunol 1980;24:177–182.

150 Melconian AK, Brun Y, Fleurette J: Enterotoxin production, phage typing and serotyping of *Staphylococcus aureus* strains isolated from clinical materials and food. J Hyg, Cambr 1983;91:235–242.

151 Holmberg SD, BLake PA: Staphylococcal food poisoning in the United States: New facts and old misconceptions. Food Poisoning 1984;251:487–489.

152 Gomez-Lucia E, Blanco JL, Goyache J, de la Fuente R, Vazquez JA, Ferri EFR, Suarez G: Growth and enterotoxin A production by *Staphylococcus aureus* S6 in Manchego type cheese. J Appl Bacteriol 1986;61:499–503.

153 Lee WH, Staples CL, Olson JC: *Staphylococcus aureus* growth and survival in macaroni dough and the persistence of enterotoxins in the dried products. J Food Sci 1975;40:119–120.

Marsha J. Betley, PhD, Department of Bacteriology, University of Wisconsin-Madison, 1550 Linden Drive, Madison, WI 53706 (USA)

Fleischer B (ed): Biological Significance of Superantigens.
Chem Immunol. Basel, Karger, 1992, vol 55, pp 36–64

T-Lymphocyte Stimulation by Microbial Superantigens

Bernhard Fleischer, Udo Hartwig

First Department of Medicine, University of Mainz, FRG

Introduction

Exotoxins of a variety of microorganisms are major pathogenicity factors. Many of these toxins are directed against cells of the defense system [1]. The staphylococcal enterotoxins (SE) are the prototypes of a group of exotoxins that have in common an extremely potent stimulatory activity for T lymphocytes of several species. These molecules have been called 'superantigens' [2] because of their efficient activation of T lymphocytes and because of the similarity of the mechanism of T-cell stimulation to antigen recognition. Recent investigations by a number of laboratories have elucidated the unusual mechanism of T-lymphocyte stimulation and have shown that at least part of the pathogenic effects of these molecules are due to their ability to stimulate a large fraction of T cells. In this review we will describe these microbial superantigens (mSA), their interaction with MHC class II and T-cell receptor (TCR) molecules and discuss molecular mechanisms of T-lymphocyte stimulation.

Microbial T-Lymphocyte Stimulating Exotoxins

Due to the popularity of 'superantigens', the list of molecules supposed to belong to this group is growing. Table 1 lists all molecules that have been proposed to be microbial superantigens (mSA). The enterotoxins and the toxic shock syndrome toxin-1 (TSST-1) of *Staphylococcus aureus* and the erythrogenic toxins A and C of *Streptococcus pyogenes* (SPE) are genetically

Table 1. Microbial proteins proposed to be 'superantigens'

Producing pathogen	Name	Abbreviation	Molecular mass kD	Status as SA
S. aureus	Enterotoxin A	SEA	27,800	established
	B	SEB	28,300	established
	C1	SEC1	26,000	established
	C2	SEC2	26,000	established
	C3	SEC3	28,900	established
	D	SED	27,300	established
	E	SEE	29,600	established
	Toxic shock syndrome toxin-1	TSST-1	22,000	established
S. aureus	Epidermolytic (exfcliative) toxin A	ExFTA	26,900	unlikely
	B		27,300	unlikely
S. pyogenes	Pyrogenic (erythrogenic) toxin A	SPEA	29,200	established
	B	SPEB	27,000	unclear
	C	SPEC	24,300	established
	M protein	pepM	22,000	unclear
M. arthritidis	Soluble mitogen	MAM	ca. 26,000	established
P. aeruginosa	Exotoxin A		66,000	unclear

related molecules of 21–30 kD [3, 4]. The SE can be distinguished by neutralizing antibodies and are divided into 5 different serotypes, of SEC there are 3 subtypes that differ only in a few amino acids [5]. It is likely that there are in addition to the known SE, TSST and the SPE other members of this group that have not yet been detected. Indeed, we have found several strains of S. aureus and S. pyogenes that produce mitogenic toxins apparently different from known toxins [unpubl. observations].

An evolutionary distant microorganism has also evolved this type of T-cell stimulation: Mycoplasma arthritidis secretes a small basic protein (MAM) that uses exactly the same molecular mechanism [6, 7]. MAM has been partially sequenced and appears to be unrelated to the SE family [6]. Discussed elsewhere in this book are the T-cell mitogens encoded by murine retroviruses that are also completely unrelated in their primary structure to the bacterial exotoxins but use a related mechanism.

Besides these established superantigens, there are other staphylococcal and streptococcal proteins completely unrelated to the enterotoxins that have been implicated to belong to the superantigen family, such as the staphylococcal epidermolytic (exfoliative) toxins [8], the streptococcal erythrogenic toxin B [9, 10] and the streptococcal M protein [11–13]. However, the mitogenicity of the exfoliative toxins and of the M protein appears to be due to contaminations with the highly potent SE or SPE [14–16] and there is a controversy about the superantigenicity of SPEB. Preparations of these proteins usually require higher concentrations for T-cell stimulation than the SE and, if used after thorough purification or in recombinant form, these molecules are not mitogenic for T cells. Because SE and SPE are active in concentrations of a few pg/ml, the sensitivity of T cells to these stimulators is several orders of magnitude higher than the sensitivity of any biochemical or serological test. To unequivocally confirm that a given microbial toxin is a superantigen, it has to be demonstrated that the isolated protein – if possible in recombinant form – is (1) active in nM concentrations; (2) specifically binds to MHC class II molecules, and (3) activates T cells in a clonally variable manner, acting on certain T-cell receptors only. It is indeed very important to use these toxins in 'physiological' concentrations and not in the μg/ml concentrations that are commonly used.

The exotoxin A of Pseudomonas aeruginosa has also been suggested to be a superantigen for mouse T cells but to require processing in contrast to all other mSA [17]. We have not been able to find mitogenicity for human T cells.

Role of Class II Molecules

Early studies already indicated a requirement for MHC class II expression on various cell types to function as accessory cells (AC) or target cells in the activation of a large proportion of T cells by staphylococcal enterotoxins A and B [18, 19]. It was also shown that SEA had to bind physically to APC to exert its mitogenic effect on T cells [20].

Since then there has been a growing body of evidence from numerous reports in the literature that MHC class II molecules are required for the presentation of microbial superantigens (mSA). Hence a dependence on class II expressing APC has been demonstrated for all superantigens studied so far, although in addition to class II molecules some other receptors for these molecules may exist (see below). In contrast, a binding of microbial superantigens to MHC class I has not been reported.

The Class II MHC Molecules are Specific Receptors for mSA

The presentation of mSA depends on accessory cells expressing class II molecules. Binding of toxins to class II positive but not negative cells has been abundantly demonstrated and this binding correlated with the ability of the cells to serve as accessory or target cells for T-cell responses to toxins [21–27]. A direct interaction between toxin and class II molecules was shown by several groups either by precipitating class II molecules with toxins bound to beads [22–24], by precipitating toxin-bound molecules with antitoxin antibodies [25] or by blotting precipitated class II molecules with labelled toxin [26]. The participation of class II molecules in the stimulation of T cells by mSA has also been demonstrated in transfection studies. Transfected but not nontransfected P815 cells could present SEA or MAM to T cells [20]. Transfection of fibroblasts with the three isotypes of HLA class II antigens (DR, DQ and DP) conferred the ability to bind SEA and to support proliferative responses at different toxin concentrations [24]. In contrast, class II negative mutants of B-cell lines that express class I only, as well as HLA class I-transfected fibroblasts, failed to induce T-cell activation with SE and TSST-1 [21–24]. The fact that Daudi cells which lack expression of MHC class I molecules, bind the mSA and stimulate CD4+ and CD8+ T cells very well [18, 25], shows that class I molecules do not play a role in the presentation of mSA in general and to CD8+ T cells in particular [18, 27]. All these findings initially made with human cells as AC were also found with murine accessory cells expressing class II antigens [2, 28–30]. Finally, the activation of T cells by exposure to mSA presented by purified IE molecules inserted into lipid bilayers [31, 32] shows that the class II antigens are not only required but also seem to be sufficient for mSA-induced T-cell stimulation, although accessory molecules can enhance this response (see below).

The capacity of purified IE molecules to induce SA-dependent T-cell activation and the observation that fixed APC can present SA as effective as untreated cells [18, 20, 35] clearly showed that the processing of SA is not required for T-cell stimulation and that these molecules act as intact proteins, not as peptides.

Whereas for conventional antigens only a single MHC allelic form is able to present a given peptide, the presentation of mSA is not MHC restricted and not even species restricted. Thus, human APC can present mSA to murine T-cell lines and T-cell hybridomas usually more efficiently than mouse APC [36]. This is for the most toxins due to a higher affinity for human as for mouse class II molecules [22, 36].

Recent reports have shown that differences exist in the capacity of different human and murine MHC isotypes to bind different mSA. It has been observed that HLA class II isotypes differ in their binding capacity for a given mSA and that different mSA have different affinities for a given class II molecule [36, 37]. Generally, the DR molecules display the highest affinity for binding mSA among the HLA isotypes [24–26]. The murine homologue of HLA-DR, the IE molecule appears to be a dominant, but (as also shown for their human homologues) not the exclusive receptor for mSA. The ability of APC from independent MHC haplotypes and intra H-2 recombinant congenic strains of mice to present SEB to murine T-cell clones revealed that in mice expressing both isotypes SEB results in the predominant presentation by IE whereas in strains lacking IE, the IA alleles differ in their ability to present SEB [38].

Interactions Between mSA and Different Allotypes and Isotypes of Class II Molecules

Although the exact configuration of the toxin-class II complex remains to be clarified, it has been reported by a number of groups that the binding of SA and conventional antigen does not occur at the same place of the MHC class II molecule, and that there exist different or overlapping binding sites for the mSA.

On the basis of the predicted peptide-binding site of the MHC class II molecule [39], an analysis of those residues of the IA^{k}α-chain proposed to form one surface of an antigen-binding groove showed that in contrast to peptides the binding and subsequent presentation of SEB was only very marginal or not at all affected by the mutations within the groove. Moreover, in competition studies SEB did not compete with the appropriate peptide for binding in the groove [40]. Furthermore, an isolated histidine to tyrosine mutation in AA81 of the DR$_\beta$-chain leads to loss of SEA binding [41]. This residue is not involved in peptide binding. From these results, it was concluded that superantigens interact with class II molecules outside of the antigen groove. This was supported by the finding that TSST-1 binding did not prevent subsequent binding of a DR-restricted antigenic peptide [42].

Among all mSA known, the possible binding sites on MHC class II molecules and the binding affinities of the staphylococcal enterotoxins are best investigated. Several approaches have been performed to identify the interaction sites between MHC class II antigens and mSA. Here we list some of the available data collected for the different toxins.

SEA is one of the most thoroughly investigated mSA. In experiments with EBV-transformed B cells or L cells transfected with HLA-DR1 it has been shown that SEA binds with high affinity ($K_d = 8.22 \times 10^{-8}$ M) to DR molecules but, although less stringent, to DQ and DP molecules [25, 37]. In addition, it has recently been suggested by competitive binding studies that two binding sites on the DR1 molecule are occupied by SEA [43], which appear to be different from or overlapping with the binding sites of SEB and TSST-1. However, the finding that a single residue determines SEA binding to DR1 [41] argues for a single binding site. An analysis of the binding sites of SEA to the murine IA isotype was performed using a synthetic peptide approach. The α-helical region 70–80 (with the residues 72, 79 or 80 being presumably most important) of the IA$_\beta^b$ chain and the region 51–80 of the IA$_\alpha^b$ chain which both are about 60% homologous with the corresponding DR sequence were identified to block SEA binding to the mouse B-cell lymphoma line A20 and to Raji cells [44, 45]. However, although the α-chain peptides directly bound SEA and inhibited T-cell activation, they were unable to inhibit this binding on APC. Thus, apparently SEA needs to contact both chains of the MHC molecule necessary for its function but the β-chain is probably sufficient for binding [45].

In addition, two domains within the SEA molecule have been suggested to contain binding sites for the MHC class II molecule. First, a peptide spanning the amino-terminal region 1-27 of SEA has been shown to bind to class II and this was proposed to be responsible for stimulation of T-cell proliferation [46]. Second, a C-terminal fragment of recombinant SEA containing amino acids 107-233 has recently been reported to bind specifically to HLA-DR, and HLA-DP, although it fails to activate human T cells [47]. Taken together, these data seem to indicate that at least two binding sites on the class II molecule and two interaction sites on the SEA molecule exist. Neither of the two cysteine residues of the toxin are required for MHC binding [48]. The stoichiometry of the interaction between SEA and the different binding sites on the class II molecule is still unclear.

SEB binds to the same or an overlapping site on HLA-DR as SEA ($K_d = 2.44 \times 10^{-7} M$) [37] as its binding can be almost completely inhibited by SEA [42]. It also binds to DQ [22]. The failure of SEB to compete efficiently with SEA for binding might be due to different or overlapping binding sites and due to the fact that SEB has a much lower affinity to DR compared to SEA. Stimulation of T cells is not supported by HLA-DP and H2-IA, and to only a low level by IE fibroblast transfectants [37]. Extensive studies on the ability of various H-2 haplotypes to present SEB to mouse T cells have been reported [38, 49]. The preservation of class II binding after loss of the disulphide loop was also shown for SEB [48].

SEC1 supports T-cell stimulation via DR and DQ isotypes and is less stimulatory potent than SEA and SEB ($K_d = 7.4 \times 10^{-7} M$).

SED is more than 70% and SEE more than 80% homologous to SEA but both toxins have a lower affinity to HLA-DR1 than the other enterotoxins studied ($1.05 \times 10^{-6} M$ and $1.11 \times 10^{-5} M$, respectively). Nevertheless, they both compete with SEA, SEB and TSST-1 binding on the DR molecule if high protein concentrations are used. In contrast to SEA, however, the binding of SED and SEE can be blocked by SEB and TSST-1. Interestingly, although SEE has the lowest affinity for DR1 of all enterotoxins, it is active to stimulate T cells in as low concentrations as SEA which has a much higher affinity [43]. Thus, the affinity of the toxin to the class II molecule does not solely determine the efficiency of T-cell stimulation.

TSST-1 has been demonstrated to bind to all 3 HLA class II isotypes but with higher affinity to HLA-DR than to DP [23, 37]. This differential affinity was used to establish (with chimeric α- and β-chains of DR and DP expressed on transfected L cells) that the α1 domain of DR is essential for high-affinity binding ($K_d = 4.4 \times 10^{-7} M$) [42]. Inhibition experiments with anti-HLA DR mAb and cross-competition with SEB on transfected L cells have strongly suggested that these two mSA bind to two different sites on HLA-DR [23, 43]. Furthermore, the ability of murine MHC class II molecules has been examined. IA+, IE– but not IA–, IE+ cells were able to support TSST-1 induced T-cell proliferation and the binding affinities seem to vary between different haplotypes [28]. L cells expressing hybrid DR_α:IE_β could present TSST-1 whereas IE_α:DR_β could not. Thus, minor differences in the highly homologous IE_α- and DR_α-chains might account for different affinities of the class II molecules for TSST-1 [28].

The erythrogenic toxin A (SPEA) of *S. pyogenes*, a further member of the enterotoxin family, does also show stimulation via the HLA-DR and HLA-DP molecules [9]. In addition, stimulation of murine T cells by SPEA in the presence of IE^k or IA^k transfected L cells has been observed [30]. Binding of SPEA cannot be blocked by TSST-1, but SPEA can inhibit binding of SEB on Raji cells [Hartwig and Fleischer, unpubl. results].

The stimulation of T cells by MAM has been shown to be absolutely dependent on the expression of a functional IE molecule whereas IA is insufficient [51]. IE molecules isolated from B cells and incorporated into liposomes as well as the transfection of the IE antigen into fibroblasts did present MAM to T cells [31]. Furthermore, L cells expressing transfected IE_α^k:IA_β^d hybrid molecules appeared to be sufficient for effective presentation of MAM since the induced proliferative response of T cells was at least equal to the response induced by AC expressing IE_α^k:IE_β^d [52]. These studies established that the T-lymphocyte proliferation in response to MAM is mediated by E_α-containing molecules and that E_β is not required for presentation. Correspondingly, on human cells the HLA-DR isotypes could be identified to interact with MAM because the presentation was inhibited by anti-HLA-DR mAb whereas anti-HLA-DP mAb did not have any effect [7]. Since MAM is not available in purified form, there is no biochemical data on its interaction with class II molecules.

Taken together, a quite complicated picture emerges from these studies. The different toxins bind to several distinguishable binding sites on class II molecules, and binding affinities depend on isotypes and allotypes. This has to be considered when the toxins and different AC are selected for T-cell stimulation assays.

The Binding of mSA to Other Receptors Different from MHC Class II Molecules

Several recent reports have indicated the existence of mSA-binding molecules different from MHC molecules. The efficient presentation of SEB and SEC1, but not SEA and SED, at picomolar concentrations by MHC class II negative colon carcinoma cell lines suggested the presence of functionally active binding structures on these cells which are different from conventional MHC class II antigens currently known [53]. Moreover, SEB, SEC1 and SEA dependent lysis of MHC class II negative target cells by murine cytolytic T cells [54] supports the idea that a non-classical or non-MHC molecule receptor may exist. The nature of this SE binding structure remains to be elucidated. There are no biochemical data available.

In addition, cells of various types have been reported to bind different toxins. Many studies have been performed with TSST-1. Porcine aortic endothelial cells, e.g., bind TSST-1 with a $K_d = 5.7 \times 10^{-7} M$ and have 2×10^4 receptors/cell [55]. TSST-1 also has been reported to bind to human umbilical vein endothelial cells [56]. Surprisingly, thrombocytes and granulocytes have also been reported to possess large numbers of receptors for the binding of SPEA [57]. The physiological significance of this finding remains to be established.

Direct Consequences of Binding to Class II Molecules

The binding of mSA to class II molecules does not only lead to a complex recognizable by T cells but also has direct consequences for the presenting cell. Apparently, this interaction can lead to the transduction of signals via class II molecules that result in the activation of class II-positive cells including monocytes, B cells, and even class II-positive activated natural killer cells and human T cells.

TSST-1 and SEB have been reported not to require the presence of T cells for their potent induction of IL-1 and TNF-α production from monocytes [48, 58]. This notion is based on experiments using human monocytes [48, 58] and the class II+ monocytic cell line THP-1 [58]. The synthesis of IL-1 and TNF-α is initiated by the transcriptional activation of the respective gene resulting in the first detection of mRNA within 30 min after TSST-1 treatment and peak levels at 1–3 h after stimulation. This transcriptional activation did not require de novo protein synthesis. Certain mAB directed to epitopes on HLA-DR which are thought to be closely related to the TSST-1 binding site also cause monokine mRNA accumulation, and indicate that TSST-1 transduces activation signals via class II molecules that result in the induction of cytokine gene transcription [58]. Further insight into the signalling by TSST-1 via class II was achieved by analysing the importance of protein kinases in the induction of cytokine gene expression using protein kinase inhibitors [59]. The results suggest an involvement of protein kinase C and of protein tyrosine kinases in the signalling pathway via class II leading to transcriptional activation of cytokine genes as this has been shown in general to be a common mechanism for the signalling of receptors.

In contrast, as discussed later, it has been shown that the stimulation of human monocytes by SEA to produce IL-1α and IL-1β [60] and TNF-α depends strictly on the presence of T cells and could not be replaced by a panel of lymphokines [61].

In human B lymphocytes, TSST-1 binding to class II molecules activates the adhesive function of the lymphocyte function-associated molecule-1 (LFA-1, CD11a/CD18) via an activation of the protein kinase C [62]. The same effect was mediated by the anti-HLA-DR mAb L243 that binds to an epitope closely related to the TSST-1 binding site. Interestingly, SEB and another anti-DR mAb failed to induce this effect. Furthermore, TSST-1 has been shown to synergize with phorbolesters and with anti-IgM in inducing B cell proliferation in the absence of T cells or T-cell factors [63]. For differentiation into Ig-secreting cells T cells were required [63]. Thus, TSST-1 delivers a comitogenic effect to B cells via MHC class II molecules as signal transducing structures.

As will be discussed below, a stimulating effect of the binding of SEA, SEB and TSST-1 to class II molecules on T cells or activated NK cells has also been reported [64].

The T-Cell Response to Microbial Superantigens

The first indication that the mechanism of T-lymphocyte stimulation by the staphylococcal enterotoxins was fundamentally different from recognition of antigen was the finding that the class II dependence of the response to SE and to MAM was found with CD4+, CD8+, and with some CD4−, CD8−, γδ+ T cells [7, 18]. Indeed, CD8+ cells can be triggered easily to destroy target cells in the presence of SE or MAM if the target cells express MHC class II molecules [7, 18, 27]. An early report that SEB could induce CTL but not trigger their cytotoxic potential was based on the use of class II negative target cells [65].

Furthermore, in contrast to conventional T-cell mitogens, such as lectins, a given toxin stimulates only a fraction of T cells. As a first hint on the mechanism of action of these toxins it was observed that human [18] and mouse [19] T-cell clones responded in a clonally variable fashion to SEA and SEB. Some clones responded only to SEA, some to SEB, and some to both. Similarly, it had been noted that MAM stimulated only a fraction of mouse T-cell hybridomas [66].

The conclusion from these observations was that variable parts of the TCR were the sites of the T cell involved. Because also γδ+ cells responded to SEA [18] such sites were apparently not restricted to αβ+ T cells. Surprisingly, the interaction with the T-cell receptor is of very low affinity because a physical binding of toxin to T cells has not been observed [20, 21].

On the αβ+ TCR, the interaction site for the toxin is the variable part of the β-chain (V_β). Incubation of mouse spleen cells [2, 35] or human peripheral blood mononuclear cells [8] with a given toxin leads to an enrichment of T cells carrying certain V_β. With several toxins such an enrichment is very strong, e.g. stimulation of human mononuclear cells with SPE-C derived from *S. pyogenes* NY5 leads to an increase of $V_\beta8+$ T cells from 4 to 30% within 5 days [15]. In fact, this stimulation of $V_\beta8+$ cells by SPE-C is so strong that minute contaminations of SPE-C appear to be responsible for the stimulation of $V_\beta8+$ T cells reported for preparations of SPE-A [9] and the streptococcal M protein [11, 13]. Injection of SEB that stimulates preferential expansion of mouse $V_\beta8$ T cells in vitro leads to a dramatic proliferation of $V_\beta8$ cells in vivo [2, 67]. If injected into neonatal mice, injection of SEB leads to clonal deletion of $V_\beta8+$ cells [2]. TSST-1 stimulates $V_\beta2+$ human T cells in vitro [68], and patients suffering from TSST-1-caused shock syndrome have elevated numbers of $V_\beta2+$ T cells in their blood [69]. Table 2 lists the associations of toxins and V_βs that have until today been reported. A problem is that many of the toxin preparations used, even the commercial ones, are contaminated by other toxins, therefore the assignments may not be correct in several cases. In addition, several toxin V_β interactions can only be seen if the toxin is presented by human class II stimulator cells [36], possibly because of the much higher affinity of human as opposed to murine class II molecules to many toxins.

The regions of the V_β contacting the toxin bound to the class II molecule has been mapped by Choi et al. [70] by genetic engineering of the TCR β-chain. SEC2 stimulates T cells carrying $V_\beta13.2$ but not T cells with $V_\beta13.1$. The exchange of the amino acids 67–78 of $V_\beta13.1$ with the amino acids of the SEC2-reactive $V_\beta13.2$ conferred reactivity to T cells carrying the engineered $V_\beta13.1$. These residues are located outside of the putative contact site with the antigenic peptide-MHC complex. Pullen et al. [71] showed, by a similar approach, that this applies also to the T-cell receptor interaction with Mls-1[a].

Table 2. Preferential stimulation of T cells carrying certain TCR V_β

Toxin	Reported preference for TCR V_β		References
	human	mouse	
SEA	3, 12, 14, 15, 17, 20	1, 3, 10, 11, 12, 17	76, 86, 114
SEB	12	7, 8, 17	2, 8, 35, 68, 76, 86
SEC1	12, 13.2, 14, 15, 17, 20	8.2, 8.3, 11, 17	8, 76, 86
SEC2	5, 12, 13.1, 13.2	8.2, 10, 17	8, 68, 70, 76, 86
SEC3	5, 8, 12	7, 8.1, 8.2, 8.3	8, 70, 76, 86
SED	5.1, 6, 6.1, 6.2, 6.3, 8, 18	7, 8, 8.1, 8.2, 8.3, 11, 17	8, 76, 86
SEE	2	11, 15, 17	8, 68, 76, 86, 115
TSST-1	8, 12, 14, 15	3, 15, 17	68, 76, 86
SPEA	1, 2, 8, 10	8.2	9, 50
SPEC	17		10, 15, 117
MAM		6, 8.1, 8.2, 8.3	6, 52, 116

The reported V_β specifications are given only for those molecules that are unequivocally established as superantigens. Discrepant results between different reports can be due to cross-contaminations with other toxins or to the existence of different subtypes that are serologically indistinguishable.

A similar selectivity as for V_β among $\alpha\beta$ TCR+ T cells has been described for $\gamma\delta$ TCR+ T cells. Among a battery of T cell clones, the reactivity to SEA presented by human EBV-transformed B cells was confined to Vγ9 bearing T cells [72]. Why only cytotoxic and no proliferative response was triggered in these clones is unclear.

How specific the interaction with a certain V_β and a given toxin is, is not completely clear. Stimulation of human T cells with, e.g., SEE leads to a preferential expansion of $V_\beta8+$ T cells and depletion of $V_\beta5+$ T cells. If, however, cloned T cells carrying $V_\beta5$ were investigated, it was found that several $V_\beta5+$ T cells could respond to SEE although such T cells were depleted from bulk cultures after stimulation with this toxin [73]. Similar findings were made with $V_\beta6+$ clones [74]. A possible explanation is that different V_βs bind to these toxins with different affinities and that T cells with the highest affinity for a given toxin are preferentially expended in bulk cultures. This notion was supported by the finding that more than 50% of human T-cell clones react with a given toxin [74]. An exception is the mitogen MAM that stimulates less than 5% of human T cells [7]. Alternatively, it has been suggested that human T-cell clones that express MHC class II molecules can be activated by binding of a toxin to the class II molecule [64, see below].

Molecular Mechanism of T-Cell Stimulation

On the basis of our own results, we have proposed [7, 18, 75] that SE and MAM are functionally bivalent molecules binding to MHC class II molecules and to variable parts of the TCR molecules present on CD4+ and CD8+ $\alpha\beta$ and on $\gamma\delta$ T cells. This idea seemed to be at that time quite unusual and was rejected by the *Journal of Immunology* [18]. Similiar, although in detail differing models have been developed by Janeway et al. [35] and Marrack and Kappler [76].

How this tri-molecular complex of toxin, T-cell receptor, and class II looks like is unclear. There are, however, a number of indications from experimental data. Basically, there are two possible mechanisms: one is that the TCR binds to a preformed complex of toxin and class II molecules but not to the isolated toxin, similar to the recognition of peptide antigen. Alternatively, there could be distinct sites on the toxin molecules binding to class II and the T-cell receptor separately. An argument for the first model could be that the toxins do not bind to the TCR. A soluble TCR-β-chain specifically bound to a complex of class II molecules and toxin but not to the isolated toxin coated to plastic [77]. However, the absence of binding does not exclude an activating effect on the T cell. A monoclonal antibody to the TCR has been described that did not bind to Jurkat cells but stimulated IL-2 production in these cells [78]. We have also found a mAb stimulating a fraction of T cells without apparent binding [79]. Possibly, multivalent low affinity interaction is sufficient to activate T cells. Another argument against a direct effect on the TCR comes from the finding that there are clone-specific interactions between toxins, T-cell receptors, and MHC molecules. Certain combinations of class II molecule and V_β do not lead to a response although the respective class II molecule is known to bind a given toxin and the respective V_β can respond to this toxin [49, 80, 86].

A number of observations argue for a direct effect of the toxins on the T-cell receptor in the absence of class II. Incubation of T cells in the absence of AC with SE, TSST-1 or MAM leads to an increase in the concentration of cytosolic Ca^{2+} [7, 18, 21, 64] and to phosphoinositol breakdown [81]. This has also been observed with class II negative T cells, e.g. with T cells from patients with class II deficiency [64]. Such a triggering of calcium release was not observed by all authors [82]. Several investigators found that isolated T cells could be stimulated by SEA or TSST in the presence of phorbol esters [18, 83], in another study, however, AC could not be replaced by activators of protein kinase C [20]. SE bound to plastic or to beads are unable to trigger T cells [20, 73]. If, however, mAb against CD2 or other T-cell surface structures are co-imobilized on the beads, a stimulation – although inefficient – results [73]. Furthermore, since class II-negative cells can apparently present SE [53, 54], the class II molecule as a presenting structure is not a prerequisite. In addition, SEA-antibody conjugates can direct CTL to kill class II negative target cells [84]. Here, SEA bound covalently to the antibody can effectively trigger CTL. Taken together, there is circumstantial evidence that a direct interaction of the toxins with the TCR takes place and results in a – subthreshold – stimulation. In this model, the subthreshold effects on the TCR are converted to a stimulatory signal by multivalent cross-linking of several TCR molecules with MHC class II molecules of the stimulating cell.

A third model of stimulation has been proposed by Janeway [85, 86]. Here the mitogenic toxin does not interact directly with V_β but with a V_β-binding endogenous 'coligand' molecule on the presenting cell. The coligand molecule binds to the superantigen class II molecule complex and this alters its affinity for self V_β-structures.

Recently, it was postulated that binding of toxins to class II molecules on T cells could deliver a stimulating signal if an additional second signal was provided [64]. This idea is based on the observation that human T-cell clones that are class II-positive, reacted to toxins in the presence of class II-negative EBV-B cells. This stimulation was apparently V_β nonspecific and did not discriminate between different toxins. The T cell was stimulated in the absence of a rise in cytosolic calcium concentration. Evidence for this type of stimulation is (1) the lack of response by class II-negative T cells, and (2) the response of CD3-TCR-negative class II-positive NK clones. Such an interaction could possibly contribute to the high percentage of human T-cell clones responding to a given toxin [73].

Requirement for Accessory Molecules

As is the case with specific recognition of antigen, the T lymphocyte uses additional adhesion molecules in the response to the superantigenic toxins. Most important appear adhesions via CD2 or LFA-1. Monoclonal antibody inhibition studies showed that these adhesions were alternatively required for the stimulation of the T cells at low toxin concentrations but that this requirement could be overcome at high concentrations [34]. Transfection of the ICAM-1 gene into fibroblasts leads to an enhanced response to SEA [33]. A costimulatory activity has also been reported for adhesion of VLA-4 to its ligand VCAM-1 [87].

Antibodies to CD4 inhibit the response of mouse T cells to SE and to Mls [19, 35]. Nevertheless, CD4-negative T-cell hybridomas respond to SE and the transfection of CD4 into these cells does not enhance their response [88]. Human T-cell responses to SE are generally not inhibited by anti-CD4 or anti-CD8 in bulk culture. Only few human CD4+ T-cell clones are weakly inhibitable by anti-CD4 [21]. Anti-CD8 does never inhibit the response of human [7, 21] or murine [27] CD8+ T cells to SE. Furthermore, CD8+ cells easily lyse MHC class I-negative cells, such as Daudi in the presence of mSA [18, 27].

Structural Requirements for Toxin Action

The SE and SPE are related molecules with approximately 230 amino acids. Some of these toxins have strong sequence homologies, e.g. SEA and SEE are more than 80% homologous in their amino acid sequences [89], whereas TSST-1 that has only 194 amino acids is hardly homologous to any

```
              1         10              20              30              40              50
SEA    * S E K S E E I N E K D L R K K S E L Q G T A L G - N L K N I Y Y - Y N E K A K T - E N K E S H D Q F L Q H
SEE        *                                    R N       S       R                I    -          D           E N
SEB      *   S Q P D P K P D E   H   S   K F T   - L M E -   M   - V L   - D D N H V S A - I   V K   I           Y F
SEC1     *   S Q P D P T P D E   H   A   K F T   - L M E -   M   - V L   - D D H Y V S A - T K V K   V   K       A
SPEA         * Q Q D P D P   Q   H R S S   V K   Q           F L   E G D P V   H     V K   V   L         S
TSST-1             * S T N D N I - K D L L D W   - S S G S D T F - T   S E V L     N S     G S


              60              70              80              90
SEA    T I L F K G F F T D H S W Y N D L L V D F D S K D I V D K Y K G K K V D L Y G - A - - - - - - Y Y G Y - Q C -
SEE    L               G   P               L G       A T N             - - - - - - - - -               -       -
SEB    D L I Y S I K D   K L G N   D N V R   E   K N     L A         D   Y     V F   - - - - - - - N   Y   -       Y
SEC1   D     Y N I S D K K L K N   D K V K T E L L N E G L A K       D E V     V     - S N Y Y V N C   F S S K D N -
SPEA   D L I Y N - - - V S G P N   D K   K T E L K N Q E M A T L F   D   N     I   V E - - - - - -       H L - C Y -
TSST-1 M - - - - - R I K N T D G S I S   I I F P S P Y Y S P A F T   D             N - - - - - - - - - T K R T - K K -


              100             110             120             130             140
SEA    - - - - - - - - - - A G G T P N K T A C M Y G G V T L H D N N R L T E E K K V P I N L W L - D G K Q N T V P L E T V
SEE    - - - - - - - - -                                                       I               T   I D K
SEB    F S K K T N D I N S H Q   D K R K T             E   N G   Q   D K Y R S I T B R V F E -       - L L S F D -
SEC1   - - - - - - - - - V   K V T G G K T           I   K   E G   H F D N G N L Q N V L I R V Y E N   R     I S F   -
SPEA   - - - - - - - - - L C E N A E R S   I           N   E G   H   E I P   I V V K V S I -     I   - S L S F D - I
TSST-1 - - - - - - - - - S N H   S E G   Y I H F Q I S G V T N T E K   P T P I E L   L K V K V - H       D S P L - K Y W P


              150             160             170             180             190
SEA    K T N K K N V T V Q E L D L Q A R R Y L Q E K Y N L Y N S D V F D G K V Q R G L I V F H T S T E P S V N
SEE        S   'E                   H       G           F G       S   D                   S   E G S T   S
SEB    Q     K   A         Y L T   H     V K N K K     E F N - - N S P Y E T   Y   K     - I E N   N   F W
SEC1   Q     S   A         I K     N F   I N   K       E F N - - S S P Y E T   Y   K     I E N N G N T F W
SPEA   E     M   A         Y K V   K     T D N K Q     T N G P - - S   Y E T   Y   K     I P K N K E   F W
TSST-1     F D   Q I A T S T     F E I   H Q   T Q I H G       R   S D - K T G G Y W K I T M N D G       Y - - - Q


              200             210             220             230
SEA    Y D L F G A Q G Q - Y S N T - L L R I Y R D N K T I N S E N M H I D I Y L Y T S *
SEE            D         -     P D   -                         L           L           T *
SEB        M M P   P   D K F D Q S K Y   M M   N           M V D   K D V K   E V       T   K K K *
SPFC1      M M P   P   D K F D Q S K Y   M M   N             V D   K S V K   E V H   T   K N G *
SPEA   F   F   P E P E F - T Q S K - Y   M     K         E   L D   N T S Q   E V       T   K *
TSST-1 S D   S K K P E Y - N T E K - P P I N I D E I       E A   I N *
```

Fig. 1. Comparison of amino acid sequences of mature SEA, SEE, SEB, SEC1, SPEA and TSST-1. Residues that a given toxin has in common with SEA are indicated by blank spaces, dashes indicate gaps. Taken from Couch et al. [89] and Betley and Mekalanos [113].

other of the toxins [3, 4]. Figure 1 shows the amino acid sequences of some of the toxins. It is apparent that there are common regions of homology in all toxins, some shared even by TSST-1. All SE and SPE with the exception of TSST-1 have a central disulphide-loop, in SEA from AA96 to AA106. In spite of their extensive homologies, SEA and SEE differ profoundly in their V_β specificities and in their binding characteristics to class II molecules [43].

The precise molecular action of the toxin molecules and the sites on the toxin molecules required for T-cell activation are still unclear. Conflicting results have been reported. Spero and Morlock [90] found that the mitogenic activity of SEC1 was confined to a 6.5-kD tryptic N-terminal fragment, and that the 22-kD C-terminal fragment was inactive. Similarly, Noscova et al. [91] found a 17-kD N-terminal fragment of SEA to be mitogenic for T cells. In contrast, Bohach et al. [92] showed that the 21-kD C-terminal tryptic fragment of SEC1 had all activities of the intact molecule. Several groups investigated the structure-function relationship of TSST-1. After fragmentation with papain, some authors found the C-terminal 12- and 16-kD fragments to be active [93], whereas others found a central 14-kD CNBr fragment (amino acids 68–159) to possess mitogenic activity [94]. Studies using modification of the central disulphide loop led to conflicting conclusions. Some authors found that the loop is not required for mitogenic activity of SEA [25], others found it important [48].

All these studies used purified protein for cleavage or modification of the toxins. A possible reason for these conflicting results is that these toxins are active in pg/ml concentrations. If biochemical experiments are performed, mg amounts are required. Minor contaminations by intact molecules, undetectable by serological or biochemical methods but active in T-cell stimulation assays, could still be present in the preparation of toxin fragments or of modified toxin. As discussed above, such contaminations have been shown to pose a major problem of work with staphylococcal proteins.

More recent studies used molecular genetic approaches to circumvent such problems. Mutations in the disulphide loop of SEA to substitute one or both of the two cysteines with different AA decreased mitogenic activity 100-fold without significantly affecting class II binding [95]. Interestingly, several mutants lost the capacity to stimulate $V_\beta11+$ cells and acquired the capacity to stimulate $V_\beta6+$ cells. The preference for $V_\beta3+$ cells was retained by all mutants. This indicates that the central disulphide-loop in SEA contributes to toxin avidity for the TCR.

Other studies used deletion mutants. C-terminal deletion of SEA led to loss of function, already if only seven amino acids had been deleted [96].

Similarly, C-terminal deletion of only 13 amino acids of the SEB molecule leads to a loss of function. A 16 amino acids deletion mutant, however, was functional if a tail of 60 amino acids encoded by the vector was added to the shortened SEB [97]. This indicates that C-terminal deletion is affecting the proper folding of the molecule and that the three-dimensional structure is important for its function. After N-terminal deletion of SEA, a fragment of AA107–233 lost mitogenicity but still bound to class II. Fragments of AA126–233 lost also class II binding. N-terminal fragments (AA1–125 and AA1–179) did neither stimulate nor had binding activity for class II [47].

A different approach to conduct a structure function analysis of TSST-1 has been the replacement of tyrosine and histidine residues by alanine [98]. Epitope scanning of the mutants by a mAb and their ability to induce T-cell stimulation resulted in the finding that specific residues near the carboxy-terminus of TSST-1 are essential for mitogenic activity and in forming the epitope recognized by a neutralizing Ab. Especially, the tyrosine residue at position 144 appears to be required for maintaining mitogenicity which has been further supported by similar analysis performed on SPEA.

The characterization of functional important regions within the SPEA toxin by generating recombinant mutants through site directed mutagenesis has revealed that certain conserved amino acids such as a tyrosine at position 160 or an arginine at position 151 appear to be crucial for the mitogenic properties of the molecule. Results from binding studies using Raji cells and a polyclonal antiserum against SPEA to detect binding by FACS analysis suggest that mutation of certain residues, e.g. 100 or 151 abrogates class II binding, whereas other mutants retain the ability to interact with class II, but lost their stimulatory activity [Hartwig and Fleischer, unpubl. results].

Recently, synthetic peptides of toxin sequences were used as functional tools. A peptide comprising the N-terminal 27 or 45 amino acids of SEA was shown to process binding activity for class II and to block mitogenicity of intact SEA [46]. Surprisingly, a synthetic 21mer peptide comprising AA58–78 of TSST-1 was recently reported to have strong mitogenic activity, to possess binding activity for class II molecules and to react with several neutralizing mAb against TSST-1 [83]. The only difference between the native TSST-1 and this peptide was the inability of the latter to stimulate purified T cells in the presence of phorbol esters [83].

Taken together, these studies still do not allow a conclusive picture of how the toxins work. Clearly, the three-dimensional molecular structure of at least one of the toxins is urgently required.

Effects of mSA in vitro

Addition of toxins to cultures of human PBMC or mouse spleen cells leads to the production of multiple cytokines by T cells and monocytes and to the vigorous proliferation of T cells both of CD4+ and CD8+ phenotype. Cytotoxic effector cells differentiate in such cultures able to lyse class II positive target cells in the presence of toxin. The toxins are extremely potent inducers of IL-1 and TNF-α in monocytes and appear to be much more effective than LPS on a molar basis [60, 61, 99, 100]. As discussed elsewhere in this book, these mediators appear to be primarily responsible for the shock-like symptoms and the pyrogenicity induced in vivo.

The production of IL-1 induced by LPS shows a different kinetic than for SEA as IL-1 can already be detected after 3–6 h and reaches maximal levels after 12 h. SEA induces a delayed release with peak values after 24–48 h [60]. SEA has also been demonstrated to be a very effective inducer of TNF-α and TNF-β production in monocytes and T cells [61] where the intracellular production of TNF-α can be detected in both cell types after 6 h. Maximal TNF activity is routinely detected after 48–72 h, and on the basis of equal frequencies of cells with intracytoplasmic TNF-α, monocytes and T cells appear to contribute to a similar extent to TNF-α found in the culture supernatant. Both the induction of IL-1 as well as TNF production in monocytes by SEA has been observed to depend strictly on the presence of T cells and could not be replaced by a panel of lymphokines such as IL-2, IL-4, IL-6, IL-7, TNF-α and interferon (IFN)-γ or combinations of those [60, 61] whereas IL-1 production by monocytes after stimulation with LPS does not depend on T cells. The T-cell requirement of SEA induced IL-1 production has been found to be mainly restricted to CD4+ 45RO+ memory T cells although it was also supported to a lesser extent by CD4+ 45RA+ naive T cells and even less by CD8+ T cells [60]. Similarly, the TNF-α production in SEA stimulated monocytes was exclusively dependent on CD4+ 45RO+ T cells [61]. Both CD8+ CD4+ T cells produced about equal amounts of TNF-β, but a fourfold higher frequency of TNF-β-producing cells was found among the CD4+ versus the CD8+ T-cell population. In addition, of the CD4+ subset, cells of both CD4+ 45RO as well as the CD4+ 45RA phenotype produced TNF-β but only CD4+ 45RA cells produced significant amounts of IFN-γ in response to SEA. Kinetic analysis demonstrated that the TNF-β production upon stimulation with SEA is much delayed compared to TNF-α as it occurs after 48 h and 6 h, respectively. The SEA-induced production of TNF-β by T cells was strongly suppressed by anti CD25 (anti-IL-2R) mAb but this mAb did not inhibit proliferation. This indicates that the TNF-β production is IL-2 dependent [61]. Those results suggest that monokine production at least in response to SEA requires T-cell contact and T-cell products derived from different cell subsets.

Incubation of cloned human T cells with suitable toxins at high concentrations (500 ng/ml) leads to a state of refractoriness or 'tolerance' that appears to differ from the refractoriness induced by antigen recognition [101]. It resembles the effects of incubation with high concentrations of specific peptide. A possible interpretation of this phenomenon is that presentation by the T cells on their class II molecule leads to an engagement of the TCR in the absence of appropriate second signals [102]. Whether this is related to the state of anergy observed in vivo is still unclear.

Binding of toxin molecules to class II molecules on B cells makes these cells recognizable for T cells bearing the appropriate TCR. This could lead in the case of cytotoxic cells to destruction of the B cell, a mechanism contributing to the immunosuppression induced by the toxins [103]. Furthermore, the interaction of helper T cells with B cells induces proliferation and differentiation of B lymphocytes into antibody-secreting plasma cells [63, 104]. This type of 'cognate' interaction circumvents the normal antigen-specific interaction and will lead to the delivery of activation signals to many B cells by a large fraction of T cell and thus might counteract mechanisms of control of autoreactive T and B cells.

Role of T-Cell Stimulation in Pathogenesis

Given the extremely low concentration of pg/ml required for T-cell stimulation and the large amounts of toxins secreted by these bacteria into the culture medium in vitro, it is clear that a small focus of bacterial infection can be sufficient to flood the immune system with toxin. Introduction of a mitogenic toxin into the body may have a number of consequences. Most importantly, as discussed in other chapters of this book, injection of a toxin induces an initial expansion of reactive T cells that is followed by death or induction of anergy in these T cells [67, 105]. In addition, the activation of cytotoxic CD8+ T cells [18, 27] could lead to destruction of antigen presenting or B cells that have bound toxin molecules to their class II antigens [103].

The shock-like symptoms induced by all of the toxins are caused by a massive release of lymphokines and monokines. The potent pyrogenicity common to all toxins is also caused by IL-1 and TNF. Although the source of TNF has not been clearly identified, this shock is dependent on the presence of T cells [106, 107]. Similar symptoms are observed after the first injection of stimulating anti-CD3 antibodies in patients requiring immunosuppressive therapy [108]. It is noteworthy that the Fc part of this antibody is required for shock induction, because F(ab)$_2$ fragments do not work. It is very surprising that TNF is already present after 1 h in the serum and that IL-2 and interferon-γ reach their maximal concentrations already after 4 h [107, 108]. Why this production occurs so much more rapidly in vivo than in vitro is unclear. Moreover, it is possible that the action of the toxins on other cells contributes to the shock induction. Recently, it has been reported that TSST-1 binds to aortic endothelial cells and induces enhanced permeability and cytotoxic effects in these cells at μg/ml concentrations [55].

The enterotoxins are known for their ability to induce a gastrointestinal illness after oral uptake of a few μg of toxin [5]. The symptoms of the gastrointestinal intoxication are apparently caused by release of leukotrienes from mast cells in mucosa [109]. Intradermal injection of SEB causes a skin reaction of anaphylactic type within a few minutes. This dermal reaction is caused by the same mediators and mechanisms as the enteric reaction [109]. Both reactions are inducible only in primates, whereas mitogenicity is found in a wider range of species [109]. The stimulation of mast cells in these reactions is indirect, probably by neuropeptides released from intramucosal ganglial cells in response to SE [110]. This would indicate that the enteropathogenic activity of SE is not caused by an action on T cells [79], but that the site of action and the receptors involved are still obscure.

Finally, these toxins could induce immunopathological phenomena. Autoreactive T cells could be nonspecifically activated, or irrelevant T cells could be focused on autoreactive T cells and induce autoantibody production. It is noteworthy in this context that for the streptococcal erythrogenic toxin A an epidemiological association with rheumatic fever has been suggested [111]. M proteins of group A streptococci carry a number of epitopes cross-reacting with human tissue [112]. Stimulation of autoreactive B cells by such epitopes and the generation of help by T cells stimulated by erythrogenic toxins may provide a pathogenetic mechanism. The requirement for such a concerted action of different pathogenicity factors could explain that autoimmune sequelae are relatively rarely found.

Concluding Remarks

The toxins described in this review constitute the most efficient T-cell stimulators known. It is conceivable that immunosuppression, destruction of antigen-presenting cells, and induction of T-cell anergy will be of advantage for the infecting microorganisms. The finding that the toxins derived from *S. aureus* and *S. pyogenes* (bacteria found only in humans) stimulate human T cells more efficiently than murine T cells, whereas the opposite is true for the mitogen derived from *M. arthritidis* (a natural pathogen for rodents), could suggest that these molecules have been adapted in evolution to the MHC and TCR molecules of the natural host. It is unlikely that endogenous superantigens exist in mice to delete T cells that would be deleterious if responding to the bacterial toxins because these bacteria are not found in wild mice. It appears that all these molecules are part of a strategy of infectious agents used to suppress the immune response by clonal deletion or anergy. That only certain TCR are addressed by a given toxin via variable regions and not all T cells via a constant region of the TCR reduces the in vivo efficacy of the toxins. However, this strategy prevents the induction of cross-reacting neutralizing antibodies against the toxins. Thus, the molecules of the enterotoxin family show a maximum of biological efficiency combined with a minimum of immunological cross-reactivity. This feature may be the reason for the generation of their polymorphisms during evolution.

Acknowledgements

Work of the authors was supported by the Bundesministerium für Forschung und Technologie.

References

1 Witholt, B. (ed.): Bacterial protein toxins (Fischer, Stuttgart 1992).
2 White, J.; Herman, A.; Pullen, A.M.; Kubo, R.; Kappler, J.W.; Marrack, P.: The V_β-specific superantigen staphylococcal enterotoxin B: Stimulation of mature T cells and clonal deletion in neonatal mice. Cell *56:* 27–35 (1989).
3 Iandolo, J.: Genetic analysis of extracellular toxins of *Staphylococcus aureus*. Annu. Rev. Microbiol. *43:* 375–402 (1989).
4 Lee, P.K.; Schlievert, P.M.: Molecular genetics of pyrogenic exotoxin 'superantigen' of group A streptococci and *Staphylococcus aureus*. Curr. Top. Microbiol. Immunol. *174:* 1–20 (1991).

5 Bergdoll, M.S.: Enterotoxin; in Easmon, Adlams, Staphylococci and staphylococcal infections, pp. 559–598 (Academic Press, New York 1983).

6 Cole, B.C.; Atkin, C.L.: The *Mycoplasma arthritidis* T cell mitogen MAM: A model superantigen. Immunol. Today *12:* 271–276 (1991).

7 Matthes, M.; Schrezenmeier, H.; Homfeld, J.; Fleischer, S.; Malissen, B.; Kirchner, H.; Fleischer, B.: Clonal analysis of human T cell activation by the mycoplasma arthritidis mitogen. Eur. J. Immunol. *18:* 1733–1737 (1988).

8 Kappler, J.; Kotzin, B.; Herron, L.; Gelfand, E.; Bigler, R.D.; Boylston, A.; Carrel, S.; Posneit, C.D.; Choi, Y.; Marrack, P.: V_β-specific stimulation of human T cells by staphylococcal toxins. Science *244:* 811–814 (1989).

9 Abe, J.; Forrester, J.; Nakahara, T.; Lafferty, J.A.; Kotzin, B.L.; Leung, D.Y.: Selective stimulation of human T cells with streptococcal erythrogenic toxins A and B. J. Immun. *146:* 3747–3750 (1991).

10 Leonard, B.A.; Lee, P.K.; Jenkins, M.K.; Schlievert, P.M.: Cell and receptor requirements for streptococcal pyrogenic exotoxin T cell mitogenicity. Infect. Immun. *59:* 1210–1214 (1991).

11 Tomai, M.; Kotb, M.; Majumdar, G.; Beachey, E.H.: Superantigenicity of Streptococcal M protein. J. exp. Med. *172:* 359–362 (1990).

12 Kotb, M.; Majumdar, G.; Tomai, M.; Beachey, E.H.: Accessory cell-independent stimulation of human T cells by streptococcal M protein superantigen. J. Immun. *145:* 1332–1336 (1990).

13 Tomai, M.A.; Aelion, J.A.; Dockter, M.E.; Majumdar, G.; Spinella, D.G.; Kotb, M.: T cell receptor V gene usage by human T cell stimulated with the superantigen streptococcal M protein. J. exp. Med. *174:* 285–288 (1991).

14 Fleischer, B.; Bailey, C.J.: Recombinant epidermolytic (exfoliative) toxin A of *Staphylococcus aureus* is not a superantigen. Med. Microbiol. Immunol. *180:* 273–278 (1991).

15 Braun, M.A.; Gerlach, D.; Hartwig, U.; Ozegowski, J.; Carrel, S.; Köhler, W.; Fleischer, B.: Stimulation of human T cells by the streptococcal 'superantigens' erythrogenic toxins A and C (scarlet fever toxins). J. Immun. (in press, 1992).

16 Fleischer, B.; Schmidt, K.H.; Gerlach, D.; Köhler, W.: Separation of mitogenic activity from streptococcal M protein. Infect. Immun. (in press, 1992).

17 Legaard, P.K.; LeGrand, R.D.; Misfeldt, M.L.: The superantigen Pseudomonas exotoxin A requires additional functions from accessory cells for T lymphocyte proliferation. Cell Immunol. *135:* 372–382 (1991).

18 Fleischer, B.; Schrezenmeier, H.: Stimulation by staphylococcal enterotoxins. Clonally variable response and requirement for MHC class II molecules on accessory or target cells. J. exp. Med. *167:* 1697–1708 (1988).

19 Janeway, C.A.; Chalupny, J.; Conrad, P.J.; Buxser, S.: An external stimulus that mimics Mls locus responses. J. Immunogenet. *15:* 161–168 (1988).

20 Carlsson, R.; Fischer, H.; Sjögren, H.O.: Binding of staphylococcal enterotoxin A to accessory cells is a requirement for its ability to activate human T cells. J. Immun. *140:* 2484–2488 (1988).

21 Fleischer, B.; Schrezenmeier, H.; Conradt, P.: T cell stimulation by staphylococcal enterotoxins. Role of class II molecules and T cell surface structures. Cell. Immunol. *119:* 92–101 (1989).

22 Herrmann, T.; Acolla, R.S.; MacDonald, H.R.: Different staphylococcal enterotox-

ins bind preferentially to distinct MHC class II isotypes. Eur. J. Immunol. *19:* 2171–2174 (1989).

23 Scholl, P.; Diez, A.; Mourad, W.; Parsonnet, J.; Geha, R.S.; Chatila, T.: Toxic shock syndrome toxin-1 binds to class II major histocompatibility molecules. Proc. natn. Acad. Sci. USA *86:* 4210–4214 (1989).

24 Mollik, J.A.; Cook, G.R.; Rich, R.G.: Class II molecules are specific receptors for staphylococcal enterotoxin A. Science *244:* 817 (1989).

25 Fraser, J.D.: High affinity binding of staphylococcal enterotoxins A and B to HLA-DR. Nature *339:* 221–223 (1989).

26 Fischer, H.; Dohlsten, M.; Lindvall, M.; Sjögren, O.; Carlsson, R.: Binding of staphylococcal enterotoxin A to HLA-DR on B cell lines. J. Immun. *142:* 3151–3157 (1989).

27 Herrmann, T.; Maryanski, J.L.; Romero, P.; Fleischer, B.; MacDonald, H.R.: Activation of MHC class I-restricted CD8+ CTL by microbial T cell mitogens. J. Immun. *144:* 1181–1186 (1990).

28 Scholl, P.R.; Sekaly, R.P.; Diez, A.; Glimcher, L.M.; Geha, R.S.: Binding of toxic shock syndrome toxin 1 to murine MHC class II molecules. Eur. J. Immunol. *20:* 1911–1916 (1990).

29 Cole, B.C.; Daynes, R.A.; Ward, J.R.: Stimulation of mouse lymphocytes by a mitogen derived from Mycoplasma arthritidis. I. Transformation is associated with an H-2-linked gene that maps to the I-E-I-C subregion. J. Immun. *127:* 1931–1936 (1981).

30 Uchiyama, T.; Saito, S.; Inoko, H.; Yan, X.J.; Imanishi, K.; Araake, M.; Igarashi, H.: Relative activities of distinct isotypes of murine and human major histocompatibility complex class II molecules in binding toxic shock syndrome toxin 1 and determination of CD antigens expressed on T cells generated upon stimulation by the toxin. Infect. Immun. *58:* 3877–3882 (1990).

31 Bekoff, M.C.; Cole, B.C.; Grey, H.M.: Studies on the mechanism of stimulation of T cells by the Mycoplasma arthritidis derived mitogen. Role of class II I-E molecules. J. Immun. *139:* 3189–3194 (1987).

32 Lee, J.M.; Watts, T.H.: Binding of staphylococcal enterotoxin A to purified murine MHC class II molecules in supported lipid bilayers. J. Immun. *145:* 3360–3366 (1990).

33 Dohlsten, M.; Hedlund, G.; Lando, P.A.; Towsdale, J.; Altmann, D.; Patavvaya, M.; Fischer, M.; Kalland, T.: Role of the adhesion molecule ICAM-1 (CD54) in staphylococcal enterotoxin-mediated cytotoxicity. Eur. J. Immunol. *21:* 131–135 (1990).

34 Mittrücker, H.W., Fleischer, B.: Stimulator cell-dependent requirement for CD2-mediated and LFA-1-mediated adhesions in T-lymphocyte activation by superantigenic toxins. Cell. Immunol. *139:* 108–117 (1992).

35 Janeway, C.A.; Yagi, J.; Conrad, P.J.; Katz, M.E.; Jones, B.; Vroegop, S.; Buxser, S.: T cell responses to Mls and bacterial proteins that mimick its behaviour. Immunol. Rev. *107:* 61–68 (1989).

36 Herman, A.; Croteau, G.; Sekaly, R.P.; Kappler, J.; Marrack, P.: HLA-DR alleles differ in their ability to present staphylococcal enterotoxins to T cells. J. exp. Med. *172:* 709–717 (1990).

37 Mollick, J.A.; Chintagumpala, M.; Cook, R.G.; Rich, R.R.: Staphylococcal exotoxin activation of T cells. Role of exotoxin-MHC class II binding affinity and class II isotype. J. Immun. *146:* 463–468 (1991).

38 Robinson, J.H.; Pyle, G.; Kehoe, M.: Influence of major histocompatibility complex haplotype on the mitogenic response of T cells to staphylococcal enterotoxin B. Infect. Immun. *59:* 3667–3672 (1991).

39 Brown, J.H.; Jardetzky, T.; Saper, M.A.; Samraoui, B.; Bjorkman, P.J.; Wiley, D.C.: A hypothetical model of the foreign antigen binding site of class II histocompatibility molecules. Nature *332:* 845–850 (1988).

40 Dellabona, P.; Peccoud, J.; Kappler, J.; Marrack, P.; Benoist, C.; Mathis, D.: Superantigens interact with MHC class II molecules outside of the antigen groove. Cell *62:* 1115–1121 (1990).

41 Herman, A.; Labrecque, N.; Thibodeau, J.; Marrack, P.; Kappler, J.W.; Sekaly, R.P.: Identification of the staphylococcal enterotoxin A superantigen binding site in the β1 domain of the human histocompatibility antigen HLA-DR. Proc. natn. Acad. Sci. USA *88:* 9954–9958 (1991).

42 Karp, D.R.; Tletski, C.L.; Scholl, P.; Geha, R.; Long, E.O.: The alpha 1 domain of the HLA-DR molecule is essential for high-affinity binding of the toxic shock syndrome toxin-1. Nature *346:* 474–476 (1990).

43 Chintagumpala, M.; Mollick, J.A.; Rich, R.R.: Staphylococcal toxins bind to different sites on HLA-DR. J. Immun. *147:* 3876–3881 (1991).

44 Pontzer, C.H.; Russell, J.K.; Johnson, H.M.: Structural basis for differential binding of staphylococcal enterotoxin A and toxic shock syndrome toxin 1 to class II major histocompatibility molecules. Proc. natn. Acad. Sci. USA *88:* 125–128 (1991).

45 Russell, J.K.; Pontzer, C.H.; Johnson, H.M.: Both alpha-helices along the major histocompatibility complex binding cleft are required for staphylococcal enterotoxin A function. Proc. natn. Acad. Sci. USA *88:* 7228–7232 (1991).

46 Pontzer, C.H.; Russell, J.K.; Johnson, H.M.: Localization of an immune functional site on staphylococcal enterotoxin A using the synthetic peptide approach. J. Immun. *143:* 280–284 (1989).

47 Hedlund, G.; Dohlsten, M.; Buell, G.; Herrmann, T.; Lando, P.A.; Segren, S.; Schrimsher, J.; MacDonald, H.R.; Sjögren, H.O.; Kalland, T.: A recombinant C-terminal fragment of SEA binds with high affinity to human MHC class II. J. Immun. *147:* 4082–4085 (1991).

48 Grossman, D.; Cook, R.D.; Sparrow, J.T.; Mollick, J.A.; Rich, R.R.: Dissociation of the stimulatory activities of staphylococcal enterotoxins for T cells and monocytes. J. exp. Med. *172:* 1831–1841 (1990).

49 Yagi, J.; Rath, S.; Janeway, C.A., Jr.: Control of T cell responses to staphylococcal enterotoxins by stimulator cell MHC class II polymorphism. J. Immun. *147:* 1398–1405 (1991).

50 Imanishi, K.; Igarashi, H.; Uchiyama, T.: Activation of murine T cells by streptococcal pyrogenic exotoxin A. Requirement for MHC class II molecules on accessory cells and identification of V_β elements in T cell receptor of toxin-reactive T cells. J. Immun. *144:* 3170–3175 (1990).

51 Cole, B.C.; Daynes, R.A.; Ward, J.R.: Stimulation of mouse lymphocytes by a mitogen derived from *Mycoplasma arthritidis.* II. Cellular requirements for T cell transformation mediated by a soluble mycoplasma mitogen. J. Immun. *128:* 2013–2018 (1981).

52 Cole, B.C.; David, C.S.; Lynch, D.H.; Kartchner, D.R.: The use of transfected fibroblasts and transgenic mice expressing E-α establishes that stimulation of $V_\beta 8$ T

cells by the *Mycoplasma arthritidis* mitogen requires E-α. J. Immun. *144:* 420–424 (1990).

53 Dohlsten, M.; Hedlund, G.; Segren, S.; Lando, P.A.; Herrmann, T.; Kelly, A.P.; Kalland, T.: Human MHC class II-colon carcinoma cells present staphylococcal superantigens to cytotoxic T lymphocytes: Evidence for a novel enterotoxin receptor. Eur. J. Immunol. *121:* 131–135 (1991).

54 Herrmann, T.; Romero, P.; Sartoris, S.; Paiola, F.; Accolla, R.S.; Maryanski, J.L.; MacDonald, H.R.: Staphylococcal enterotoxin-dependent lysis of MHC class II negative target cells by cytolytic T lymphocytes. J. Immun. *146:* 2504–2512 (1991).

55 Lee, P.K.; Vercellotti, G.M.; Deringer, J.R.; Schlievert, P.M.: Effects of staphylococcal toxic shock syndrome toxin 1 and aortic endothelial cells. J. infect. Dis. *164:* 711–719 (1991).

56 Kushnaryov, V.M.; MacDonald, H.S.; Reiser, R.F.; Bergdoll, M.S.: Reaction of toxic shock syndrome toxin I with endothelium of human umbilical cord vein. Rev. infect. Dis. *11:* suppl., pp. 282–288 (1989).

57 Scriba, S.; Wagner, B.; Wagner, M.: Receptors for erythrogenic toxin A of S. pyogenes on human peripheral blood lymphocytes and other blood cells. Acta histochem., suppl. XXXIII, pp. 17–22 (1986).

58 Trede, N.; Geha, R.S.; Chatila, T.: Transcriptional activation of monokine genes by MHC class II ligands. J. Immun. *146:* 2310–2315 (1991).

59 Chatila, T.; Scholl, P.; Spertini, F.; Ramesh, N.; Trede, N.; Fuleihan, R.; Geha, R.S.: Toxic shock syndrome toxin-1, toxic shock and the immune system. Curr. Top. Microbiol. Immunol. *174:* 63–80 (1991).

60 Gjörloff, A.; Fischer, H.; Hedlund, G.; Hansson, J.; Kenney, J.S.; Allison, A.C.; Sjögren, H.O.; Dohlsten, M.: Optimal induction of interleukin-1 in human monocytes by the superantigen staphylococcal enterotoxin A requires the participation of T helper cells. Cell. Immunol. *137:* 61–71 (1991).

61 Fischer, H.; Dohlsten, M.; Andersson, U.; Hedlund, G.; Ericsson, P.O.; Hansson, J.; Sjögren, H.O.: Production of TNF-α and TNF-β by staphylococcal enterotoxin A activated human T cells. J. Immun. *144:* 4663–4669 (1990).

62 Mourad, W.; Geha, R.S.; Chatila, T.: Engagement of major histocompatibility complex class II molecules induces sustained, LFA-1 dependent cell adhesion. J. exp. Med. *172:* 1513–1516 (1990).

63 Fuleihan, R.; Mourad, W.; Geha, R.S.; Chatila, T.: Engagement of MHC-class II molecules by the staphylococcal exotoxin TSST-1 delivers a progression signal to mitogen activated B cells. J. Immun. *146:* 1661–1666 (1991).

64 Spertini, F.; Spits, H.; Geha, R.S.: Staphylococcal exotoxins deliver activation signals to human T cell clones via major histocompatibility complex class II molecules. Proc. natn. Acad. Sci. USA *88:* 7533–7538 (1991).

65 Zehavi-Willner, T.; Berke, G.: The mitogenic activity of staphylococcal enterotoxin B: A monovalent T cell mitogen that stimulates cytolytic T lymphocytes but cannot mediate the lytic interaction. J. Immun. *137:* 2682 (1986).

66 Lynch, D.H.; Cole, B.C.; Bluestone, J.A.; Hodes, R.: Cross-reactive recognition by antigen-specific, MHC-restricted T cells of a mitogen derived from *Mycoplasma arthritidis* is clonally expressed and I-E restricted. Eur. J. Immunol. *16:* 747–751 (1986).

67 MacDonald, H.R.; Baschieri, S.; Lees, R.K.: Clonal expansion precedes anergy and death of $V_\beta 8^+$ peripheral T cells responding to staphylococcal enterotoxin B in vivo. Eur. J. Immunol. *21:*1963–1966 (1991).

68 Choi, Y.; Kotzin, B.; Herron, L.; Callahan, J.; Marrack, P.; Kappler, J.: Interaction of staphylococcal aureus toxin superantigens with human T cells. Proc. natn. Acad. Sci. USA 86: 8941–8945 (1989).

69 Choi, Y.; Lafferty, J.A.; Clements, J.R.; Todd, J.K.; Gelfand, E.W.; Kappler, J.; Marrack, P.; Kotzin, B.L.: Selective expansion of T cells expressing $V_\beta 2$ in toxic shock syndrome. J. exp. Med. 172: 981–984 (1990).

70 Choi, Y.; Herman, A.; DiGiusto, D.; Wade, T.; Marrack, P.; Kappler, J.: Residues of the variable region of the T cell receptor β-chain that interact with S. aureus toxin superantigens. Nature 346: 471–473 (1990).

71 Pullen, A.M.; Wade, T.; Marrack, P.; Kappler, J.W.: Identification of the region of T cell receptor β-chain that interacts with the self-superantigen Mls-1a. Cell 29:1365–1374 (1990).

72 Rust, C.J.; Verreck, F.; Vietor, H.; Koning, F.: Specific recognition of staphylococcal enterotoxin A by human T cells bearing receptors with the $V_\beta 9$ region. Nature 346: 572–574 (1990).

73 Fleischer, B.; Gerardy-Schahn, R.; Metzroth, B.; Carrel, S.; Gerlach, D.; Köhler, W.: A conserved mechanism of T cell stimulation by microbial toxins. Evidence for different affinities of T cell receptor-toxin interaction. J. Immun. 146:11–17 (1991).

74 Goronzy, J.J.; Oppitz, U.; Weyand, C.M.: Clonal heterogeneity of superantigen reactivity in human $V_\beta 6^+$ T cell clones: Limited contributions of V_β sequence polymorphisms. J. Immun. 148: 604–611 (1992).

75 Fleischer, B.; Schrezenmeier, H.: Staphylococcal enterotoxins: MHC class II-dependent probes for the T cell antigen receptor. Immunobiology 175: 328–329 (1987).

76 Marrack, P.; Kappler, J.: The staphylococcal enterotoxins and their relatives. Science 248: 705–711 (1990).

77 Gascoigne, N.R.J.; Ames, K.T.: Direct binding of secreted T cell receptor β chain to superantigen associated with class II major histocompatibility complex protein. Proc. natn. Acad. Sci. USA 88: 613–616 (1991).

78 Carrel, S.; Giuffre, L.; Vacca, A.; Salvi, S.; Mach, J.P.; Isler, P.: Monoclonal antibodies against idiotypic determinants of the T cell receptor from HPB-ALL induce IL-2 production in Jurkat cells without apparant evidence of binding. Eur. J. Immunol. 16: 823 (1986).

79 Alber, G.; Scheuber, P.H.; Reck, B.; Sailer-Kramer, B.; Hartman, A.; Hammer, D.K.: Role of substance P in immediate-type skin reactions induced by staphylococcal enterotoxin B in unsensitized monkeys. J. Allergy clin. Immunol. 84: 880–885 (1989).

80 Fleischer, B.; Mittrücker, H.W.: Evidence for T cell receptor-HLA class II molecules interaction in the response to superantigenic bacterial toxins. Eur. J. Immunol. 21: 1331–1333 (1991).

81 Chatila, T.; Wood, N.; Parsonnet, J.; Geha, R.S.: Toxic shock syndrome toxin-1 induces inositol phospholipid turnover, protein kinase C translocation and calcium mobilization in human T cells. J. Immun. 140: 1250–1255 (1988).

82 Qasim, W.; Kehoe, M.A.; Robinson, J.H.: Does staphylococcal enterotoxin B bind directly to murine T cells? Immunology 73: 433–437 (1991).

83 Edwin, C.; Swack, J.A.; Williams, K.; Boventre, P.F.; Kass, E.H.: Activation of in vitro proliferation of human T cells by a synthetic peptide of toxic shock syndrome toxin 1. J. infect. Dis. 163: 524–529 (1991).

84 Dohlsten, M.; Hedlund, G.; Akerblom, E.; Lando, P.A.; Kalland, T.: Monoclonal antibody-targeted superantigens: A different class of anti-tumor agents. Proc. natn. Acad. Sci. USA *88:* 9287–9291 (1991).

85 Janeway, C.A.: Self superantigens. Cell *63:* 659–661 (1990).

86 Janeway, C.A., Jr.: Selective elements for the V_β region of the T cell receptor: Mls and the bacterial toxic mitogens. Adv. Immunol. *50:* 1–53 (1991).

87 van Seventer, G.A.; Newman, W.; Shimizu, Y.; Nutman, T.B.; Tanaka, Y.; Horgan, K.J.; Gopal, T.V.; Ennis, E.; O'Sullivan, D.; Grey, H.; Shaw, S.: Analysis of T cell stimulation by superantigen plus major histocompatibility complex class II molecules or by CD3 monoclonal antibody: Costimulation by purified adhesion ligands VCAM-1, ICAM-1 but not ELAM-1. J. exp. Med. *174:* 901–913 (1991).

88 Sekaly, R.P.; Croteau, G.; Bowman, M.; Scholl, P.; Burakoff, S.; Geha, R.S.: The CD4 molecule is not always required for the T cell response to bacterial enterotoxins. J. exp. Med. *173*: 367–371 (1991).

89 Couch, J.L.; Soltis, M.T.; Betley, M.J.: Cloning and nucleotide sequence of the type E staphylococcal enterotoxin gene. J. Bact. *170:* 2954–2960 (1988).

90 Spero, L.; Morlock, B.A.: Biological activities of the peptides of staphylococcal enterotoxin C formed by limited tryptic hydrolysis. J. biol. Chem. *253:* 8787–8791 (1978).

91 Noskova, V.P.; Ezepchuk, Y.V.; Nosko, A.N.: Topology of the functions in molecule of staphylococcal enterotoxin type A. Int. J. Biochem. *1984:* 201–206.

92 Bohach, G.A.; Handley, J.P.; Schlievert, P.M.: Biological and immunological properties of the carboxyl terminus of staphylococcal enterotoxin C1. Infect. Immun. *57:* 23–28 (1989).

93 Edwin, C.; Kass, E.H.: Identification of functional antigenic segments of toxic shock syndrome toxin 1 by differential immunoreactivity and by differential mitogenic responses of human peripheral blood mononuclear cells, using active toxin fragments. Infect. Immun. *57:* 2230–2236 (1989).

94 Blomster-Hautamaa, D.A.; Novick, R.P.; Schlievert, P.M.: Localization of biological function of toxic shock syndrome toxin-1 by use of monoclonal antibodies and cyanogen bromide generated toxin fragment. J. Immun. *137:* 3572–3576 (1986).

95 Grossmann, D.; Van, M.; Mollick, J.A.; Highlander, S.K.; Rich, R.R.: Mutation of the disulphide loop in staphylococcal enterotoxin A: Consequences for T cell recognition. J. Immun. *147:* 3274–3281 (1991).

96 Hufnagle, W.O.; Tremaine, M.T.; Betley, M.J.: The carboxyl-terminal region of staphylococcal enterotoxin type A is required for a fully active molecule. Infect. Immun. *59:* 2126–2134 (1991).

97 Metzroth, B.; Linnig, M.; Fleischer, B.: Effect of deletions on function and antigenicity of staphylococcal enterotoxin B (submitted, 1992).

98 Blanco, L.; Choi, E.M.; Connolly, K.; Thompson, M.R.; Bonventre, P.F.: Mutants of staphylococcal toxic shock syndrome toxin-1: Mitogenicity and recognition by a neutralizing monoclonal antibody. Infect. Immun. *58:* 3020–3028 (1990).

99 Parsonnet, J.: Mediators in the pathogenesis of toxic shock syndrome: An overview. Rev. infect. Dis. *11:* suppl., pp. 263–269 (1989).

100 Fast, D.J.; Schlievert, P.M.; Nelson, R.D.: Toxic shock syndrome-associated staphylococcal and streptococcal pyrogenic toxins are potent inducers of tumor necrosis factor production. Infect. Immun. *57:* 291–294 (1989).

101 O'Hehir, R.E.; Lamb, J.R.: Induction of specific clonal anergy in human T lymphocytes by S. aureus enterotoxins. Proc. natn. Acad. Sci. USA *87:* 8884–8889 (1990).

102 Schwartz, R.M.: A cell culture model for T lymphocyte clonal anergy. Science *248:* 1349–1355 (1990).

103 Dohlsten, M.; Hedlund, G.; Kalland, T.: Staphylococcal enterotoxin-dependent cell-mediated cytotoxicity. Immunol. Today *12:* 147–150 (1991).

104 Mourad, W.; Scholl, P.; Diez, A.; Geha, R.; Chatila, T.: The staphylococcal toxin TSST-1 triggers B cell proliferation and differentiation via MHC unrestricted cognate T/B cell interaction. J. exp. Med. *170:* 2011–2022 (1989).

105 Kawabe, Y.; Ochi, A.: Programmed cell death and extrathymic reduction of $V_\beta 8^+$ CD4$^+$ T cells in mice tolerant to *Staphylococcus aureus* enterotoxin B. Nature *349:* 245–248 (1991).

106 Marrack, P.; Blackman, M.; Kushner, E.; Kappler, J.: The toxicity of staphylococcal enterotoxin B in mice is mediated by T cells. J. exp. Med. *171:* 455–464 (1990).

107 Miethke, T.; Wahl, C.; Heeg, K.; Echtenacher, B.; Kramer, P.H.; Wagner, H.: T cell-mediated lethal shock triggered in mice by the superantigen staphylococcal enterotoxin B: critical role of tumor necrosis factor. J. exp. Med. *175:* 91–98 (1992).

108 Chatenoud, L.; Ferran, C.; Bach, J.F.: The anti-CD3 induced syndrome: a consequence of massive in vivo cell activation. Curr. Top. Microbiol. Immunol. *174:* 121–134 (1991).

109 Scheuber, P.H.; Golecki, J.R.; Kickhöfen, B.; Scheel, D.; Beck, G.; Hammer, D.K.: Skin reactivity of unsensitized monkeys upon challenge with staphylococcal enterotoxin B: A new approach for investigating the site of toxin action. Infect. Immun. *50:* 869–876 (1985).

110 Alber, G.; Hammer, D.K.; Fleischer, B.: Relationship between enterotoxic and T lymphocyte stimulating activity of staphylococcal enterotoxin B. J. Immun. *144:* 4501–4506 (1990).

111 Yu, C.; Ferretti, J.: Molecular epidemiologic analysis of the type A streptococcal exotoxin in clinical *S. pyogenes* strains. Infect. Immun. *57:* 3715–3719 (1989).

112 Fischetti, V.A.: Streptococcal M protein: Molecular design and biological behaviour. Clin. Microbiol. Rev. *2:* 285–314 (1989).

113 Betley, M.J.; Mekalanos, J.J.: Nucleotide sequence of the type A staphylococcal enterotoxin gene. J. Bact. *170:* 34–41 (1988).

114 Takimoto, H.; Yoshikai, Y.; Kishihara, K.; Matsuzaki, G.; Kuga, H.; Otani, T.; Nomoto, K.: Stimulation of all T cells bearing $V_\beta 1$, $V_\beta 3$, $V_\beta 11$ and $V_\beta 12$ by a staphylococcal enterotoxin A. Eur. J. Immunol. *20:* 617–621 (1990).

115 Uchiyama, T.; Yan, X.J.; Imanishi, K.; Kawachi, A.; Araake, M.; Tachihara, R.: Activation of murine T cells by staphylococcal enterotoxin E: requirement of MHC class II molecules expressed on accessory cells and identification of V_β sequence of T cell receptors in T cells reactive to the toxin. Cell. Immunol. *133:* 446–455 (1991).

116 Friedman, S.M.; Crow, M.K.; Tumang, J.R.; Tumang, M.; Yiqing Xu; Hodtsev, A.S.; Cole, B.C.; Posnett, D.N.: Characterization of human T cells reactive with the Mycoplasma arthritidis-derived superantigen (MAM): Generation of a monoclonal antibody against $V_\beta 17$, the T cell receptor gene product expressed by a large fraction of MAM-reactive human T cells. J. exp. Med. *174:* 891–900 (1991).

117 Tomai, M.A.; Schlievert, P.M.; Kotb, M.: Distinct T cell receptor V_β gene usage by human T lymphocytes stimulated with the streptococcal pyrogenic exotoxins and pepM5 protein. Infect. Immun. *60:* 701–705 (1992).

B. Fleischer, MD, First Department of Medicine, University of Mainz, Langenbeckstrasse 1, D-W-6500 Mainz 1 (FRG)

Fleischer B (ed): Biological Significance of Superantigens.
Chem Immunol. Basel, Karger, 1992, vol 55, pp 65–86

Retroviral Superantigens

Hans Acha-Orbea[1]

Ludwig Institute for Cancer Research, Epalinges, Switzerland

Introduction

The immune system has evolved a near unlimited repertoire of receptor structures to recognize most of the foreign antigens. The specificity of a cellular immune response is given by the T-cell receptor (TCR) molecules expressed on the cell surface of a given T-cell clone. Each T-cell clone expresses one type of the highly polymorphic TCR molecules. In addition, many accessory molecules are involved in augmenting the overall affinity but not the specificity of these T cells. The CD4 molecules have been shown to be expressed by the T-cell subset interacting with major histocompatibility complex (MHC) class II molecules, whereas the CD8 molecules are expressed by T cells interacting with MHC class I molecules (cytotoxic T cells) [1]. These TCR molecules are α/β chain heterodimers. This enormous TCR repertoire is generated by somatic recombination of the different TCR elements during maturation of the T lymphocytes, a process similar to the generation of the vast antibody repertoire. In the germ line, clusters of variable (50–100 Vα, 25 Vβ), diversity (2 Dβ), junctional (12 Jβ, 50–100 Jα) and constant (1 Cα, 2 Cβ) regions are encoded in the mouse. During recombination, nucleotides are deleted or inserted at random in the VD, DJ and VJ junctions. These mechanisms allow the immune system to have a potential repertoire of TCR of in the order of 10^{15} different molecules [2].

Foreign antigens are recognized by T cells in the context of self-MHC molecules. It has been recently shown that T cells recognize processed proteins (peptide fragments) bound to the polymorphic MHC molecules [for review, see ref. 3]. About $1/10^5$ peripheral T cells is specific for a given MHC-

[1] Supported by a START fellowship of the Swiss National Science Foundation.

peptide complex and all the polymorphic components of the TCR (Vβ, Vα, Dβ, Jα, Jβ and the highly polymorphic region at the junction of these elements) are important for this form of antigen recognition. A frequency of 0.5–5% of T cells is reactive with allo-MHC molecules. Part of this reactivity is most likely directed at different peptides bound to the allo-MHC molecules.

Minor lymphocyte-stimulating (Mls) antigens have been found which induce a strong mixed lymphocyte reaction (MLR) between MHC-identical strains of mice [4]. Several different such genes have been discovered, and the classical 4 independently segregating loci were named *Mls-1, -2, -3* and *-4*. Each of these loci encodes a stimulatory allele (e.g. *Mls-1ᵃ*) and a nonstimulatory 'allele' (e.g. *Mls-1ᵇ*).

Contrary to classical antigens, the number of Mls-reactive cells is in the order of 1/3–1/20. The classical Mls-presenting cell is the B cell. Other antigen-presenting cells (APCs), such as macrophages, are incapable of efficiently presenting Mls antigens to T cells. Recent experiments indicated that T cells, especially CD8⁺ T cells, are able to present Mls as well [5, 6]. In addition, similar to bacterial enterotoxins (see other reviews in this volume), MHC class II expression is required for Mls presentation [for review, see ref. 7]. This became clear when it was shown that there is a hierarchy in Mls presentation in MHC class II congenic mouse strains [8].

Up to now it was impossible to raise antibodies against Mls or to get any information about the biochemical or molecular nature of Mls antigens. The only assay system was the stimulation of Mls-reactive T cells with Mls-expressing B cells in vitro.

It was shown that the T cells interacting with one particular Mls determinant express the same TCR Vβ chain. Mls-1ᵃ, for example, interacts with T cells expressing Vβ6, 7, 8.1 and 9, which make up over 20% of peripheral CD4⁺8⁻ or CD4⁻8⁺ single-positive T cells in most Mls-1ᵇ mouse strains [9–12]. Therefore, only one portion of the highly polymorphic TCR molecules is sufficient for interaction of T cells with Mls-MHC complexes. In mice expressing Mls antigens, the reactive T cells are not present in the peripheral repertoire due to deletion during maturation of the T cells.

Very recently, however, light was shed on the nature of these endogenous superantigens. Most of them are closely linked to endogenous mouse mammary tumor viruses (MMTVs) and the viral protein responsible for this effect has been characterized [13–20; for review, see ref. 21].

Mls antigens have been key tools for our current understanding of immunological tolerance (self-nonself discrimination). Knowledge of the endogenous superantigens opens new strategies for refining these mechanisms.

This review will summarize the biological effects of endogenous superantigens, describe the characterization of the viral proteins responsible for Mls action and finally discuss the biological relevance of these findings.

Tolerance

Mls antigens and TCR transgenic mice have been key tools for our current understanding of tolerance mechanisms. T cells enter the thymus as double-negative CD4⁻CD8⁻ precursor cells, differentiate to double-positive and then into the mature single-positive (CD4⁺CD8⁻ or CD4⁻CD8⁺) T cells which are then exported into the periphery and make up the peripheral T-cell repertoire. During this process, T cells which are interacting too strongly with self-MHC (and self-peptide?) are eliminated by a process called negative selection, and T cells with a weak recognition of self-MHC molecules are selected for export to the periphery by a mechanism called positive selection. In mice expressing Mls antigens, the Mls-reactive T cells are deleted by negative selection in the thymus most likely due to self-reactivity. Therefore, Mls-1ᵃ-expressing mice, for example, express very few T cells expressing TCR Vβ6, 7, 8.1 and 9 in the periphery. Surprisingly, both the CD4⁺ as well as the CD8⁺ subset are deleted although Mls is presented exclusively by MHC class II molecules [10].

Deletion of self-reactive cells can also be observed in the periphery [22, 23]. Other peripheral tolerance mechanisms have been described which lead to unresponsiveness (anergy) of T cells [23–26]. Such T cells are no longer able to proliferate in vitro in the absence of interleukin-2 (IL-2), do not produce IL-2 and cannot be stimulated by antigen.

Mice expressing Mls antigens from birth delete the majority of Mls-reactive T cells from their peripheral repertoire [9, 10]. When Mls-expressing B or T cells are injected into neonatal mice, deletion of the reactive T cells occurs with similar kinetics as in mice expressing these Mls determinants endogenously [23, 27]. When Mls-1ᵃ-expressing T or B cells are injected into adult mice, the reactive T cells can either proliferate and increase to up to 40% of the peripheral repertoire 4 days after injection, or not increase at all [23, 24; Acha-Orbea, in preparation]. A proliferative response is dependent on the mouse strains used for the experiment and on the type of Mls antigen used. Even in the absence of proliferation, however, the reactive T cells are rendered anergic about 2 days after injection [23, 24]. These results are very similar to recent experiments with the bacterial superantigen, staphylococcal enterotoxin B [25, 26].

Immunological Properties of Mls

Strong MLRs between MHC-compatible strains of mice can be generated using Mls-expressing stimulator B cells and responder T cells from Mls-disparate strains of mice [4; for review, see ref. 7]. Until the discovery of thymic deletion of T cells expressing particular TCR Vβ chains, T cell stimulation was the only biological assay system to define Mls phenotypes. It proved impossible to raise Mls-specific antibodies, and any attempts to characterize the molecular nature of Mls antigens failed. Planar membranes of Mls-expressing cells or fixation of these Mls-presenting cells destroyed their Mls presentation capability even though such preparations were able to stimulate peptide-specific or alloreactive T cells [Acha-Orbea, unpubl. data]. Irradiation of the Mls-presenting cells as well abolished Mls presentation in an irradiation-dose-dependent manner [Acha-Orbea, unpubl. data]. Addition of biochemical extracts of Mls-expressing cells to Mls-negative APCs did not lead to Mls presentation even after incorporation of the membrane proteins into liposomes. Injection of biochemical extracts of Mls-1^a-expressing lymphoma cells into Mls-1^b neonatal mice did not lead to deletion of the reactive T cells [Acha-Orbea and J. Tschopp, unpubl. data]. Injection of serum of Mls-1^a-expressing mice into newborn Mls-1^b mice was used to see whether Mls activity is present in the serum of mice or can be liberated after destruction of the highly radiation-sensitive B cells. Serum of mice irradiated lethally 24 h beforehand but not serum of unirradiated mice resulted in a reproducible partial deletion of Vβ6-expressing T cells (from 11 to 9%). After biochemical fractionation of the serum components, however, the deleting activity was lost. We had indications that the deleting capacity was retained in very labile high-molecular-weight complexes (≫100kDa) [Acha-Orbea and M. Peitsch, unpubl. data].

It was clear from the beginning that MHC class II molecules are responsible for presentation of Mls antigens to T cells. Amongst the two MHC class II isotypes I-A and I-E, I-E is in general the better-presenting molecule than I-A and, amongst the I-A molecules, I-A^k and I-A^b are good, I-A^d, I-A^s and I-A^f are intermediate presenters, and I-A^q is not able to present the different Mls genes at all [8, 9, 28]. In addition, mutagenesis experiments replacing single amino acids in the I-Aα chain of MHC class II molecules clearly indicated that the polymorphic amino acid determining the classical peptide specificity of binding to MHC class II did not play a major role in Mls interaction. On the other hand, amino acid replacements on the side of the α-helices of the MHC class II sequences diminished Mls presentation partially [29].

With the observation that Mls antigens are able to delete the reactive T cells during thymic maturation (see above), new Mls-like determinants were found. In this review I call them Mls-like since they differ from the classical Mls determinants in the following parameters.

(1) They do not induce a strong MLR in most strain combinations. Only a minor proportion of T cell clones and hybridomas expressing the relevant TCR molecules are able to react with Mls-like determinants in vitro. An exception is found in the C58 strain of mice which can be stimulated by Mls-like determinants interacting with TCR Vβ5, 11 and 12 [30].

(2) Mls-like determinants can only be presented in the context of I-E but not in the context of I-A for deletion in vivo [18, 31–39].

(3) Deletion in vivo is generally slower with Mls-like determinants (several weeks to months after birth for complete deletion) than with Mls gene products (3–10 days after birth) [18, 27].

B cells are the best Mls-presenting cells in vitro [40, 41]. In vivo injection showed that 20–50 times fewer CD8$^+$ T cells are required for induction of deletion than B cells [6]. Whether T cells are able to present Mls antigens themselves or whether an in vitro or in vivo transfer of Mls to APCs happens is not clear at present. The MHC class II molecules which are required for Mls presentation are absent or very low on mouse T lymphocytes.

MHC-H-2^q-expressing APCs are not able to present Mls [28]. An in vitro presentation of Mls antigens was observed, however, after co-incubation of Mls-1^a-expressing H-2^q and Mls-1^b-expressing H-2^k APCs [28]. Similar experiments suggested a transfer of Mls antigens in vivo. Bone marrow chimeras between H-2^q, Mls-1^a irradiated hosts and H-2^d, Mls-1^b donor bone marrow cells resulted in mice deleting T cells expressing TCR Vβ6, 7, 8.1 and 9 [42, 43].

Mls antigens have been implicated in induction of suppressor cells [for review, see ref. 6]. Most likely these effects were due to the above-described induction of anergy of the Mls-responsive T cells.

Induction of Mls-Reactive Cytotoxic T Cells

Since MHC-class-II-reactive T cells in general express CD4 molecules, it was surprising to find that in Mls-expressing mice both the CD4$^+$ as well as the CD8[+] T cells are deleted in the thymus. These results were interpreted as indications that negative selection occurs at the CD4[+]CD8[+] (double-positive) stage of maturation [10]. When anti-Mls primary MLRs are analyzed for the responsive T-cell subsets it is clear that in T-cell populations containing both subsets the CD4[+] T cells respond preferentially, but a highly significant increase of CD8[+] Vβ6[+] T cells can be found in the blast population. Up to 60% of the reactive CD8[+] cells express Vβ6. Although these T cells are functionally efficient cytotoxic T cells, they are not able to lyse Mls-expressing target cells [Acha-Orbea, unpubl. data]. Whether important second signals are lacking or whether the low density of antigen on the APCs allows stimulation but not cytotoxicity is currently not known. Cytotoxic T-cell clones have been described which can be stimulated to proliferation by Mls-1[a]-expressing target cells, but no cytotoxic response could be measured with these T-cell clones [for review, see ref. 7].

Genetic Evidence That Mls Could Be Encoded by
Endogenous Copies of Mouse Mammary Tumor Virus

DBA/2 (H-2[d], Mls-1[a]) mice were crossed and backcrossed with BALB/c (H-2[d], Mls-1[b]) mice and a congenic BALB/c strain, BALB.D2 (H-2[d], Mls-1[a]), was derived [44]. Using this mouse strain, it was possible to map the Mls-1 gene to chromosome 1 [44]. It is known that irradiation of mice, like any treatment which activates DNA repair mechanisms, leads to activation of endogenous retroviruses. Based on the biochemical serum fractionation experiments, we decided to reevaluate a role for retroviruses in Mls presentation. In the region of chromosome 1 where Mls-1[a] is localized, an endogenous copy of MMTV is encoded *(Mtv-7)* [44]. Results from the literature clearly indicate that there is no correlation between Mls-1[a] phenotype and *Mtv-7* genotype. We retyped a large panel of mouse strains with known Mls-1 phenotype for Mtv content and found that all the strains which did not fit the Mls/*Mtv-7* correlation found in the literature were mistyped. A perfect correlation between Mls expression and *Mtv-7* genotype was found. In the same analysis we observed a perfect correlation between *Mtv-6* genotype and Mls-3[a] phenotype [18]. Independently, three other groups made very similar observations for a total of 11 Mls and Mls-like determinants [13–17, 19, 20, 45]. The characterization of Mls and Mls-like antigens and the endogenous or exogenous MMTVs encoding them are summarized in table 1.

Table 1. Characterization of TCR Vβ deletion elements (Mls and Mls-like antigens)

Nomenclature	MLR	Vβ deleted	Retrovirus	Chromosome	MHC class II	References
Mls-1ᵃ, Mlsᵃ	+	6, 7, 8.1, 9	*Mtv-7* ORF	1	E > A	18, 46, Huber et al., submitted
Mls-2ᵃ, Mlsᶜ	?	3, 17	*Mtv-13*	4	E ≫ A	46
Mls-3ᵃ, Mlsᶜ	+	3, 5, 17	*Mtv-6* ORF	16	E ≫ A	18, 20, 46
Mls-4ᵃ, Mlsᶜ	?	3	*Mtv-1*	7	E > A	46
Mlsᶜ	?	3, 17	*Mtv-3*	11	E > A	15
Etc-1, Dvb11.2	-	5.1, 5.2, 11	*Mtv-9* ORF	12	E	13, 16, 17
Dvb11.1	-	11	*Mtv-8*	6	E	16
Dvb11.3	-	11	*Mtv-11*	14	E	16
-	-	14	*Mtv-2*	18	E	18
-	-	14	MMTV (GR) ORF	exogenous	E	18
-	-	14, 15	MMTV (C3H) ORF	exogenous	E	14, 19, W.H. and H.A.-O., unpubl.
-	-	12	?	?	E	35, 37, 38
-	-	16	?	?	E	35, 37, 38
-	-	17a	?	?	E	31, 32
-	-	19a	?	?	E	Hodes, unpubl.
-	-	20	?	?	E	39
-	?	5	MuLV?	exogenous?	A	79

Note that the TCR Vβ specificities are not complete yet.

Mouse Mammary Tumor Virus

Each laboratory mouse strain and most wild mouse strains have integrated MMTV copies. In each laboratory mouse, 2–8 copies of these proviral loci are found. A total of 43 different integrations have been mapped so far, and the loci were named *Mtv-1, -2, -3* ... [46]. These proviruses are stably integrated into the germ line and are inherited like normal mouse genes. Characteristic bands for each locus are generated by Southern blot analysis after restriction endonuclease digestion and probing with MMTV probes. Most of these proviruses have been mapped to specific chromosomal localizations in the mouse genome. The nucleotide and amino acid sequences of these proviruses are in the order of 95% homologous. As the name MMTV says, these retroviruses are responsible for a large proportion of mammary tumors in female mice [for review, see ref. 47]. From the endogenous proviruses, *Mtv-1* and *Mtv-2* have retained the capability of inducing infectious virus particles and of inducing mammary tumors [48]. For unknown reasons, the other endogenous viruses have lost this capacity or are possibly able to induce tumors in very old mice only.

Like all the known integrated retroviruses, MMTVs are composed of a 5' and a 3' long terminal repeat (LTR) containing sequences important for the regulation of expression. Between the two LTRs of the MMTV proviruses, group-specific antigen *(gag)*, polymerase *(pol)* and envelope *(env)* genes are encoded. A schematic diagram of the genomic organization of MMTVs, the important mRNA molecules and proteins detectable in MMTV-expressing cells are shown in figure 1 [for review, see ref. 48]. Characteristic for MMTV are the large LTR regions. They contain an open reading frame (ORF) which extends in the 3' LTR from the end of the *env* gene for about 960 nucleotides. MMTV RNA is expressed at highest levels in the mammary gland and at lower levels in lymphocytes. After stimulation of B lymphocytes, the level of MMTV RNA increases drastically [49, 50]. The biological function of ORF molecules is not known. It has been suggested that the ORF gene products play a role in positive or negative gene regulation [51, 52]. In the N-terminal half of the molecule, 5 potential ATG start codons can be found. It is not known which of the start codons is utilized by the majority of ORF gene products in vivo. It is possible that the different gene products fulfil different functions. After the second ATG a hydrophobic sequence of 25 amino acids is present which could serve as either a signal peptide (if the second ATG is used) or as a transmembrane region (if the first ATG is used). In the N-terminal half of the molecule, 5 potential N-linked glycosylation sites are

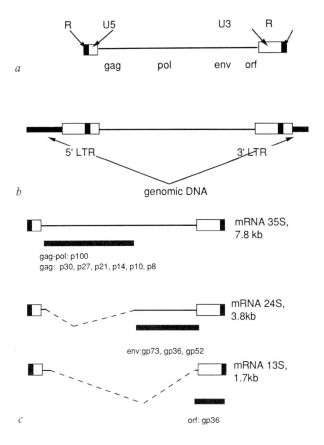

Fig. 1. MMTV, proteins and its RNA. *a* cDNA of MMTV formed after infection. R, U3, U5. *b* MMTV after integration into the mouse genome. *c* Relevant mRNA molecules and major proteins.

present. After expression in insect cells, different glycosylated and nonglycosylated forms can be detected [53].

The ancestor of these proviruses is an infectious MMTV B-type particle which can be found in mouse milk of several mouse strains at high titers ($> 10^{12}$ virus particles/ml of milk). It has been shown that the immune system, instead of protecting the animal, plays an important role in tumor development. Nude mice which do not develop a thymus, and therefore express very few T lymphocytes, are not susceptible to tumor development after uptake of exogenous virus. After injection of T cells from infected mice, however, these mice develop mammary tumors [54]. It is thought that the immune system serves as a reservoir for MMTVs until the mammary gland gets receptive for infection. Since the infection occurs at birth, neonatal tolerance is induced towards MMTVs and the immune system is incapable of destroying the infected mammary cells. After infection of the mammary epithelial cells, infectious virus particles are produced and shed into the milk and integration close to oncogenes *(int-1* to *int-5)* mammary tumors can develop [55]. During the first week of life the stomach is not acidified and infectious virus can pass to the small intestine. There the virus is taken up, most likely through the M cells, and most likely enters Peyer's patches. After infection through breast-feeding, the immune system is infected and the vicious circle is closed. Tumors usually develop between 6 and 12 months with the exogenous MMTVs and later with the endogenous copies. In aged mice, tumors could develop either after activation of endogenous viruses or by mechanisms independent of MMTV. It has been shown that treatment of mice or cultured mouse cells with irradiation or carcinogenic agents can cause liberation of normally not infectious endogenous viruses and lead to formation of infectious viral particles [56–59]. Common to these agents is their ability to induce DNA repair mechanisms.

In rare cases, MMTVs can integrate into the germ line which is represented by the endogenous proviruses found in different mouse populations. *Mtv-2* is the most recent example of acquisition of a new provirus.

Mls Are Encoded by Mouse Mammary Tumor Viruses

To directly prove a role of MMTVs in the effects of Mls on the immune system, we tested transgenic mice expressing the entire MMTV genome found in GR mice. This virus is called MMTV(GR). It is the infectious particle which gave rise to *Mtv-2*. This provirus is not expressed in any of the common laboratory mouse strains. No results are available concerning a role for this virus in deletion of T cells. Therefore, the finding that any MMTV has the capacity of deleting T cells expressing specific TCR Vβ chains would strongly indicate that practically all the known Mls and Mls-like antigens are encoded by MMTV. These transgenic mice were tested for deletion of T cells expressing any of the TCR Vβ chains. Monoclonal antibodies specific for most mouse TCR Vβ regions are available. The results of this analysis are shown in figure 2. Mice expressing the transgene show a time-dependent slow deletion of T cells expressing TCR Vβ14 as compared to transgenic litter-

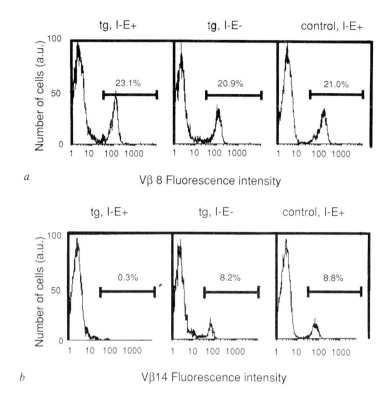

Fig. 2. Clonal deletion in MMTV(GR) transgenic mice. Transgenic mice expressing the entire MMTV(GR) genome were tested for TCR expression. *a* Expression of CD4$^+$ Vβ8$^+$ T cells in lymph nodes of 6-month-old animals. *b* Expression of CD4$^+$ Vβ14$^+$ T cells. Transgenic mice expressing MHC class II I-E are shown in the left, transgenic mice not expressing I-E in the middle, and nontransgenic I-E-expressing littermates in the right panels.

mates. Deletion is not complete until 36 weeks after birth. In addition, I-E expression is required for deletion to occur. Furthermore, no strong MLRs can be observed with transgenic stimulator and nontransgenic responder cells. Therefore, MMTV(GR) behaves like an Mls-like determinant [18].

In the strain background used for these experiments, Vβ14 is expressed only by a very small proportion of CD8$^+$ T cells due to lack of positive selection [60]. Therefore, we generated F$_1$ mice between transgenic mice and mice expressing the H-2^k MHC haplotype, which results in high-level expression of TCR Vβ14 in the CD8$^+$ subset of T cells due to positive selection.

Table 2. Deletion of CD4 and CD8 T cells expressing Vβ14 in MMTV(GR) transgenic mice

	Vβ14/CD4⁺	Vβ6/CD4⁺	Vβ14/CD8⁺	Vβ6/CD8⁺
Transgenic	0.3	9.3	2.0	13.3
Nontransgenic	6.5	9.6	10.0	12.5

Transgenic mice, which express the H-2^d MHC haplotype, were crossed with B10.Br (H-2^k) to increase expression of TCR Vβ14 in the CD8⁺ subpopulation. At 6 months of age the mice were tested for TCR expression by FACS analysis.

As shown in table 2, Vβ14⁺ CD8⁺ T cells are deleted at slower kinetics than Vβ14⁺ CD4⁺ T cells.

To further characterize the MMTV gene responsible for the Mls-like activity, we tested transgenic mice carrying the full-length ORF gene driven by a heterologous promoter. Like the entire MMTV, this construct as well led to deletion of T cells expressing Vβ14. Therefore, the Mls-like activity could be clearly mapped to the ORF molecule [18].

At the same time it was shown that uptake of milk of C3H mice leads to deletion of T cells expressing Vβ14 [14]. Choi et al. [19] showed in transfection experiments that this activity as well is localized in the ORF molecule of MMTV (C3H). Very recently it was shown that ORF molecules are responsible for several other Mls and Mls-like effects [20; Huber et al., submitted] (see table 1).

At the C terminus of ORF molecules a highly polymorphic region can be found [18, 19, 61–66]. When the ORF molecules of endogenous or exogenous MMTVs are lined up, a striking correlation between the amino acid sequence at its C terminus and the TCR Vβ chains recognized can be observed (fig. 3). The other regions which differ between the ORF molecules show no correlation with the TCR specificities. To prove the interpretation that the C-terminal part of ORF molecules is involved in TCR interaction, we tested transgenic mice carrying a variant of MMTV (C3H) isolated from an adenocarcinoma of the kidney. This variant has changed the C-terminal part of the ORF molecule completely. These transgenic mice did not show any indications for deletion of T cells expressing TCR Vβ14 [18].

Additional transgenic mice were tested which contained the gene encoding the MMTV(GR) ORF molecule starting at the fifth in frame ATG, in the middle of the molecule. This molecule contains the polymorphic ORF C

```
                    250      260      270      280      290      300      310      320
         TCR Vβ deleted |        |        |        |        |        |        |        |
MTV-8      11      PFRERLARARPPWCVLTQEEKDDIKQQVHDYIYLGTGMNVWGKIFHYTKEGAVARQLEHISADTFGMSYNG
MTV-9      5,11    ----------------------M----------------------------L----------------
MTV-11     11      ---------------FS------M---------------------------------------------

MTV-6      5,3     -----------------------M---------------IH-*-V-YNSR-E-KRHII---K-LPLAF
MTV-1      3       ---------------S-------------------*IH-*-V-YNSR-E-KRHII---K-LP

C3H        14      --------------M-S------M---------------HF------*----T--GLI--Y--K-Y----YE
GR         14      --------------M--EK----M---------------HF---V--*----T--GLI--Y--K-Y----YD

BR-6       ?(14?)  S------------M------N---------------SSI------*---RT--ALI--Y--K-Y---YD
MTV-17     ?       -----------------K-----E---------R-RI---K---RCI----------------DIR--I
GR-K       ?       ---------------S------M-------V-R----RDLNVF-K*SR-EVQKHLID**-IKALPL--
```

Fig. 3. Sequence homologies. The published ORF amino acid sequences are listed. The sequences start at position 250 and extend to the C terminus. They are grouped in ORF molecules with similar TCR specificity. Note that the TCR specificities within a group are not complete yet.

terminus which is implicated in TCR recognition. Like the full-length ORF transgenic mice, this gene was driven by a heterologous promoter. These transgenic mice did not show any deletion of T cells expressing Vβ14. Although it is tempting to speculate that half of the ORF molecule is not enough for induction of deletion, other interpretations cannot be ruled out. It is possible that this gene product is not stable, not properly folded or simply not transported to the relevant cellular compartments. Expression of the relevant proteins, addition to Mls-negative APCs and analysis of the T cell responses are required to finally settle this point.

Comparison between Retroviral and Bacterial Superantigens

The bacterial superantigens have numerous biological similarities with Mls and Mls-like antigens. There are, however, several striking differences. A comparison between the two classes of superantigens is given in table 3 [for review, see ref. 21 and 67]. Bacterial superantigens are described in more details in other chapters of this volume. Using the bacterial superantigens, a strong cytotoxic response towards MHC-class-II-expressing target cells can be observed [68, 69]. Different bacterial superantigens show different MHC class II isotype specificity. Class-II-expressing cells from a wide variety of

Table 3. Comparison of bacterial and retroviral superantigens

	Bacterial superantigens	Retroviral superantigens
Stimulation	+	+ or –
Deletion	+	+
Anergy	+	+
Cytotoxicity	+	–
APCs B cells[a]	+	+
Class-II-expressing transfectants	+	–
Macrophages	+	–
Sequence homology to host proteins	none	none
Sequence homology between members[b]	20–80%	95%
Binding to MHC class II	strong	? (most likely)
Binding to TCR	? (most likely)	? (most likely)
Processing required for presentation	no	?

[a] Effectivity of B cells as Mls-presenting cells.
[b] Amino acid sequence homology between different bacterial and retroviral superantigens.

species can be utilized as stimulators or as targets in cytotoxicity assays. In addition, there is no preference for CD4$^+$ T cells over CD8$^+$ T cells to proliferate in response to these superantigens. Any class-II-expressing cell, even transfected fibroblasts [70], can serve as stimulator cells. This is in striking contrast to the properties of the endogenous superantigens. The reasons for these differences are not understood yet. The bacterial superantigens show a much lower amino acid sequence homology than MMTV ORF molecules. The different members of bacterial superantigens fall into several families which have 20–80% amino acid sequence homology [67]. On the other hand, MMTV ORF molecules are about 95% homologous and most of the polymorphisms are located at their C terminus.

Mechanisms

With the informations available about the two classes of superantigens, a hypothetical model is shown in figure 4. Analysis of TCR Vβ chains which have lost their capability of interaction with particular superantigens which were either found in wild mouse populations or generated by site-directed

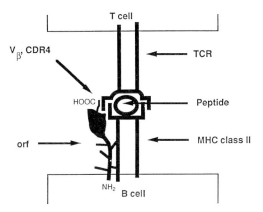

Fig. 4. Hypothetical model for Mls-TCR-MHC class II interaction. The ORF molecule is drawn as a type II molecule, inserted into the membrane through its N terminus. The C terminus of the ORF molecule is drawn as a loop. The 5 glycosylation sites are shown in the stem of the molecule.

mutagenesis indicated that the CDR4 region of the TCR molecule is important for interaction with either class of superantigen [71–74]. Mutation of single amino acids can abolish or regenerate this interaction. Hypothetical model structures of the TCR based on its homology with immunoglobulin molecules indicated that this part of the TCR molecule is situated on a side of the TCR molecule far away from the proposed peptide and MHC interaction sites. For bacterial superantigens, strong binding to MHC class II molecules was shown [75]. Single amino acid substitutions in the N-terminal part of the class II I-Aα chain showed that specific changes in the peptide-binding groove lead to loss of peptide recognition by T cells but leave superantigen recognition unaffected. Changes on the side of the model structure, however, reduced the efficiency of bacterial superantigen presentation [29]. Therefore, it seems likely that superantigens cross-link TCR and MHC class II molecules. This mechanism, however, has not been directly proven yet.

Evolution

Superantigens have been found in bacteria and retroviruses. Although they have many features in common, these proteins have very little sequence homology. Different classes of bacterial superantigens can be grouped and the different members of a group are about 80% homologous in amino acid sequence. Between groups, however, the homology can be as low as 20%. No sequence homology can be found between retroviral and bacterial superantigens. In addition, no host genes (besides the integrated members of the MMTV family) show any significant homology to these superantigens. Why have both bacteria and retroviruses developed or kept proteins which can strongly influence the immune response? Most likely these proteins are an advantage for the parasites. Experiments indicated that an overstimulation of the immune system by superantigens can lead to immune paralysis instead of an immune response [23–26, 76]. With the description of anergy and peripheral deletion of superantigen-reactive T cells, this immune suppression becomes understandable.

In the case of retroviruses, two interpretations for the importance of superantigens have been proposed.

On one hand, it has been argued that mice have not eliminated the endogenous superantigens because expression of them prevents reinfection with similar retroviruses or protects the mice from autoimmunity by deleting autoreactive T cells [67, 77]. Along the same lines, Janeway et al. [78] suggested that Mls-expressing mice are able to mount a more efficient immune response.

On the other hand, it has been proposed that expression of endogenous superantigens does not represent a disadvantage for the mouse [21]. Deletion of T cells expressing several TCR Vβ chains seems not to decrease fitness of the mice since wild mice have been found which delete a large proportion of their TCR Vβ genes by deletion in the germ line and nevertheless are able to form large colonies. Since mammary tumors develop only late in life and most endogenous MMTVs have lost the capacity of inducing mammary tumors efficiently, tumor development is unlikely to be important for natural selection. Using this interpretation, *Mtvs* would represent fossils of evolution. Since no selection against their expression occurs, mutation rates would not be higher than for host gene intron sequences. It is not known how long the known *Mtvs* have been integrated into the mouse genome. Rough estimates are in the range of 1–5 million years, a rather short time in an evolutionary scale. It is also possible that these integrations occurred much later. Analysis of the pattern of *Mtv* expression in wild mouse colonies which have been separated for defined times could help clarify these questions.

Retroviruses have an enormous growth advantage over mammalians. Their growth rate allows them to reach densities of about 10^{12} viral particles per milliliter of milk. Reverse transcriptase is an enzyme which makes more errors in translating RNA into DNA than the host DNA copying machinery. These two facts together make it likely that the retroviruses have adopted this strategy to their advantage and not to the advantage of the host.

Infection at birth most likely leads to neonatal tolerance towards the virus. Therefore, it seems unlikely that MMTV has chosen this strategy to evade an immune response. One possibility is that the efficiency of infection of the virus is not very high. This is reflected in the difficulty to infect mouse cells in vitro. After successful infection, insertion is facilitated by stimulation of T and B cells, and the infected cells most likely propagate and amplify the number of infected cells. The inability of the host to mount a cytotoxic response guarantees survival of the integrated virus.

Immunology

Superantigens have been key elements for our current understanding of tolerance mechanisms. With the finding that MMTV encodes such superantigens, tools have become available to explore the immune system, tumor immunology and retrovirology even further. The advantage of this system is that Mls antigens can be used as infectious agents. Injection with MMTV leads to lifelong infection of the mice.

We have started investigation of the immune response against MMTV using known and new exogenous MMTVs. We showed that different types of MMTVs lead to different forms of stimulation of the immune cells, and we will use these approaches in the near future in some of the following experiments.

Careful analysis of the early immune responses towards ORF proteins will allow an accurate comparison of neonatal and adult immune responses.

Local injection of MMTV with viruses affecting the immune response differently will allow to study the mechanisms of tolerance in vivo in detail.

Autoimmune diseases are caused by autoreactive lymphocytes. For at least one of these diseases a limited repertoire of TCR of the autoimmune T cells was proven. Injection of the different exogenous viruses which are able to delete different TCR-Vβ-chain-expressing T cells for the whole life of the infected mice will allow to show directly whether particular TCR Vβ chains are implicated in the initiation of autoimmunity in any of the mouse models of autoimmune disease. These experiments require a single injection with the retrovirus and using the different retroviruses yields a well-controlled experimental system.

The types of cells which are infected in the early phases after injection can then be determined.

References

1 Bierer BE, Sleckman BP, Ratnofsky SE, Burakoff SJ: The biological roles of CD2, CD4 and CD8 in T-cell activation. Annu Rev Immunol 1989;7:578–599.
2 Davis MM, Bjorkman PJ: T-cell antigen receptor genes and T-cell recognition. Nature 1988;334:395–402.
3 Rothbard JB, Gefter ML: Interactions between immunogenic peptides and MHC proteins. Annu Rev Immunol 1991;9:527–565.
4 Festenstein H: Immunogenic and biological aspects of in vitro allotransformation (MLR) in the mouse. Transplant Rev 1973;15:62–88.
5 Larsson-Sciard EL, Spetz-Hatgberg AL, Casrouge A, Kourilisky P: Analysis of T cell receptor V_β gene usage in primary mixed lymphocyte reactions: Evidence for directive usage by different antigen-presenting cells and Mls-like determinants on T cell blasts. Eur J Immunol 1990;20:1223–1229.
6 Webb SR, Sprent J: Induction of neonatal tolerance to Mls[a] antigens by CD8[+] T cells. Science 1990;248:1643–1646.
7 Abe R, Hodes R: T cell recognition of minor lymphocyte stimulating (Mls) gene products. Annu Rev Immunol 1989;7:683–708.
8 MacDonald HR, Glasebrook AL, Schneider R, Lees RK, Pircher H, Pedrazzini T, Kanagawa O, Nicolas J-F, Howe RC, Zinkernagel RM, Hengartner H: T-cell reactivity and tolerance to Mls[a]-encoded antigens. Immunol Rev 1989;107:89–108.
9 Kappler JW, Staerz UD, White J, Marrack PC: Self-tolerance eliminates T cells specific for Mls-modified products of the major histocompatibility complex. Nature 1988;332:35–40.
10 MacDonald HR, Schneider R, Lees RL, Howe RK, Acha-Orbea H, Festenstein H, Zinkernagel RM, Hengartner H: T-cell receptor Vβ use predicts reactivity and tolerance to Mls[a]-encoded antigens. Nature 1988;332:40–45.
11 Happ MP, Woodland DC, Palmer E: A third T cell receptor V_β gene encodes reactivity to Mls-1[a] gene products. Proc Natl Acad Sci USA 1989;86:6293–6296.
12 Okada CY, Holzmann B, Guidos S, Palmer E, Weissman IL: Characterization of a rat antibody specific for a determinant encoded by the $V_\beta 7$ gene segment. J Immunol 1990;144:3473–3477.
13 Woodland D, Happ MP, Bill J, Palmer E: Requirement for cotolerogenic gene products in the clonal deletion of I-E reactive T cells. Science 1990;247:964–967.
14 Marrack P, Kushnir E, Kappler J: A maternally inherited superantigen encoded by a mammary tumor virus. Nature 1991;349:524.
15 Fairchild S, Knight AM, Dyson J, Tomonari K: Co-segregation of a gene encoding a deletion ligand for Tcrb-V3[+] T cells with Mtv-3. Immunogenetics, in press.
16 Dyson PJ, Knight AM, Fairchild S, Simpson E, Tomonari K: Genes encoding ligands for deletion of Vβ11 T cells cosegregate with mammary tumor virus genomes. Nature 1991;349:531–532.
17 Woodland DL, Happ MP, Gollub KJ, Palmer E: An endogenous retrovirus mediating deletion of αβ T cells? Nature 1991;349:529–530.

18 Acha-Orbea H, Shakhov AN, Scarpellino L, Kolb E, Müller V, Vessaz-Shaw A, Fuchs R, Blöchlinger K, Rollini P, Billote J, Sarafidou M, MacDonald HR, Diggelmann H: Clonal deletion of $V_\beta 14$ positive T cells in mammary tumor virus transgenic mice. Nature 1991;350:207–211.

19 Choi Y, Kappler JW, Marrack P: A superantigen encoded in the open reading frame of the 3' long terminal repeat of mouse mammary tumor virus. Nature 1991;350: 203–207.

20 Woodland DL, Lund FE, Happ MP, Blackman MA, Palmer E, Corley RB: Endogenous superantigen expression is controlled by mouse mammary tumor proviral loci. J Exp Med 1991;174:1255–1258.

21 Acha-Orbea H, Palmer E: Mls – A retrovirus exploits the immune system. Immunol Today 1991;12:356–361.

22 Jones LA, Chin LT, Longo DL, Kruisbeek AM: Peripheral clonal elimination of functional T cells. Science 1990;250:1726–1729.

23 Webb S, Morris C, Sprent J: Extrathymic tolerance of mature T cells: Clonal elimination as a consequence of immunity. Cell 1990;63:1249–1256.

24 Rammensee H-G, Kroschewsky R, Frangoulis B: Clonal anergy induced in mature $V_\beta 6$ T lymphocytes on immunizing Mls-1^b mice with Mls-1^a expressing cells. Nature 1989;339:541–544.

25 Kawabe Y, Ochi A: Programmed cell death and extrathymic reduction of $V_\beta 8^+$ CD4$^+$ T cells in mice tolerant to *Staphylococcus aureus* enterotoxin B. Nature 1991;349: 245–248.

26 MacDonald HR, Baschieri S, Lees RK: Clonal expansion precedes anergy and death of $V_\beta 8^+$ peripheral T cells responding to staphylococcal enterotoxin B in vivo. Eur J Immunol 1991;21:1963–1966.

27 Schneider R, Lees RK, Pedrazzini T, Zinkernagel RM, Hengartner H, MacDonald HR: Postnatal disappearance of self-reactive ($V_\beta 6^+$) cells from the thymus of Mls^a mice: Implications for T cell development and autoimmunity. J Exp Med 1989;169: 2149–2158.

28 DeCruiff R, Ju S, Laning J, Cantor H, Dorf M: Activation requirements of cloned inducer T cells. III. Need for two stimulator cells in the response of a cloned line to Mls determinants. J Immunol 1986;137:1109–1114.

29 Dellabonna P, Peccaud J, Kappler J, Marrack P, Benoist C, Mathis D: Superantigens interact with MHC class II molecules outside of the antigen groove. Cell 1990;62:1115–1121.

30 Abe R, Kanagawa O, Sheard MA, Malissen B, Foo-Phillips M: Characterization of a new minor lymphocyte stimulatory system. I. Cluster of self antigens recognized by 'I-E-reactive' V_βs, $V_\beta 5$, $V_\beta 11$, and $V_\beta 12$ T cell receptors for antigen. J Immunol 1991;147:739–749.

31 Kappler JW, Wade T, White J, Kushnir W, Blackman M, Bill J, Roehm N, Marrack P: A T cell receptor V_β segment that imparts reactivity to a class II major histocompatibility complex product. Cell 1987;49:263–271.

32 Kappler JW, Roehm N, Marrack P: T cell tolerance by clonal elimination in the thymus. Cell 1987;49:273–280.

33 Tomonari K, Lovering E: T-cell receptor-specific monoclonal antibodies against a $V_\beta 11$-positive mouse T cell clone. Immunogenetics 1988;28:445–451.

34 Bill J, Kanagawa O, Woodland D, Palmer E: The MHC molecule I-E is necessary but not sufficient for the clonal deletion of $V_\beta 11$-bearing T cells. J Exp Med 1989;169: 1405–1419.

35 Vacchio MS, Hodes R: Selective decrease in T cell receptor V_β expression: Decreased expression of specific V_β families is associated with expression of multiple MHC and non-MHC gene products. J Exp Med 1989;170:1335–1346.

36 Okada CY, Weissman IL: Relative V_β transcript levels in thymus and peripheral lymphoid tissues from various mouse strains: Inverse correlation of I-E and Mls expression with relative abundance of several V_β transcripts in peripheral lymphoid tissues. J Exp Med 1989;169:1703–1719.

37 Vacchio MS, Ryan JJ, Hodes RJ: Characterization of the ligand(s) responsible for negative selection of $V_\beta 11$- and $V_\beta 12$-expressing T cells: Effects of a new Mls-determinant. J Exp Med 1990;172:807–813.

38 Singer PA, Balderas RS, Theofilopoulos AN: Thymic selection defines multiple T cell receptor V_β 'repertoire phenotypes' at the CD4/CD8 subtype level. EMBO J 1990;9:3641–3648.

39 Six A, Jouvin-Marche E, Loh DY, Cazenave PA, Marche PN: Identification of a T cell receptor β chain variable region, $V_\beta 20$, that is differentially expressed in various strains of mice. J Exp Med 1991;174:1263–1266.

40 Webb SR, Okamoto A, Ron Y, Sprent J: Restricted tissue distribution of Mls[a] determinants. J Exp Med 1989;169:1–12.

41 Molina IJ, Cannon NA, Hyman R, Huber B: Macrophages and T cells do not express Mls-1[a] determinants. J Immunol 1989;143:39–44.

42 Pullen AM, Marrack P, Kappler JW: The T-cell repertoire is heavily influenced by tolerance to polymorphic self-antigens. Nature 1988;335:796–801.

43 Speiser DE, Schneider R, Hengartner H, MacDonald HR, Zinkernagel RM: Clonal deletion of self-reactive T cells in irradiation bone marrow chimeras and neonatally tolerant mice: Evidence for intercellular transfer of Mls[a]. J Exp Med 1989;170:595–600.

44 Festenstein H, Berumen L: BALB.D2-Mls[a] – A new congenic mouse strain. Transplantation 1983;37:322–324.

45 Frankel WN, Rudy C, Coffin JM, Huber BT: Linkage of Mls genes to endogenous mammary tumor viruses of inbred mice. Nature 1991;349:526–528.

46 Kozak C, Peters G, Pauley R, Morris V, Michaelides R, Dudley J, Green M, Davisson M, Prakash O, Vaidya A, Hilgers J, Verstraeten A, Hynes N, Diggelmann H, Peterson D, Cohen JC, Dickson C, Sarkar N, Nusse R, Varmus H: A standardized nomenclature for endogenous mouse mammary tumor viruses. J Virol 1987;61: 1651–1654.

47 Bentvelzen P, Hilgers J: Murine mammary tumor virus; in Klein G (ed): Viral Oncology. New York, Raven Press, 1980, pp 311–355.

48 Dickson C: Molecular aspects of mouse mammary tumor virus biology. Int Rev Cytol 1985;108:119–147.

49 King LB, Lund FE, White DA, Sharma S, Corley RB: Molecular events in B lymphocyte differentiation: Inducible expression of the endogenous mouse mammary tumor proviral gene, Mtv-9. J Immunol 1990;144:3218–3227.

50 King LB, Corley RB: Lipopolysaccharide and dexamethasone induce mouse mammary tumor proviral gene expression and differentiation in B lymphocytes through distinct pathways. Mol Cell Biol 1990;10:4211–4220.

51 van Klaveren P, Bentvelzen P: Transactivating potential of the 3' open reading frame of murine mammary tumor virus. J Virol 1988;62:4410–4413.

52 Salmons B, Erfle V, Brem G, Günzburg WH: naf, a trans-regulating negative-acting factor encoded within the mouse mammary tumor virus open reading frame region. J Virol 1990;64:6355–6359.

53   Brandt-Carlson C, Butel JS: Detection and characterization of a glycoprotein encoded by the mouse mammary tumor virus long terminal repeat gene. J Virol 1991;65:6051–6060.

54   Tsubura A, Inaba M, Imai S, Murakami A, Oyaizu N, Yasumizu R, Ohnishi Y, Tanaka H, Morii S, Ikehara S: Intervention of T-cells in transportation of mouse mammary tumor virus (milk factor) to mammary gland cells in vivo. Cancer Res 1988;48:6555–6558.

55   Peters G, Brookes S, Smith R, Placzek M, Dickson C: The mouse homolog of the hst/k-FGF gene is adjacent to int-2 and is activated by proviral insertion in some virally induced mammary tumors. Proc Natl Acad Sci USA 1989;86:5678–5682.

56   Timmermans A, Bentvelzen P, Hageman PC, Calafat J: Activation of mammary tumor virus in 020 strain mice by X-irradiation and urethane. J Gen Virol 1969; 4:619–621.

57   Boot LM, Bentvelzen P, Calafat J, Röpcke G, Timmermans A: Interaction of X-ray treatment, a chemical carcinogen, hormones and virus in mammary gland carcinogenesis. Oncology 1970;1:434–443.

58   Links J, Calafat J, Buijs F, Tol O: Simultaneous chemical induction of MTV and MLV in vitro. Eur J Cancer 1977;13:577–587.

59   Ruppert B, Wei W, Medina D, Heppner GH: Effect of chemical carcinogen treatment on the immunogenicity of mouse mammary tumors arising from alveolar nodule outgrowth lines. J Natl Cancer Inst 1978;61:1165–1169.

60   Liao N-S, Maltzman J, Raulet DH: Positive selection determines T cell receptor $V_\beta 14$ gene usage by CD8[+] T cells. J Exp Med 1989;170:135–143.

61   Donehower LA, Huang AL, Hagler GL: Regulatory and coding potential of the mouse mammary tumor virus long terminal repeat. J Virol 1981;37:226–237.

62   Fasel N, Pearson K, Buetti E, Diggelmann H: The region of mouse mammary tumor virus DNA containing the long terminal repeat includes a long coding sequence and signals for hormonally regulated transcription. EMBO J 1982;1:3–7.

63   Majors JE, Varmus HE: Nucleotide sequencing of an apparent proviral copy of env mRNA defines determinants of expression of the mouse mammary tumor virus env gene. J Virol 1983;47:495–504.

64   Donehower LA, Fleurdelys B, Hagler GL: Further evidence for the protein coding potential of the mouse mammary tumor virus long terminal repeat: Nucleotide sequence of an endogenous proviral long terminal repeat. J Virol 1983;45:941–949.

65   Crouse C, Pauley RJ: Molecular cloning and sequencing of the Mtv-1 LTR. Virus Res 1989;12:123–138.

66   Moore R, Dixon M, Smith R, Peters G, Dickson C: Complete nucleotide sequence of a milk-transmitted mouse mammary tumor virus: Two frameshift suppression events are required for translation of gag and pol. J Virol 1987;61:480–490.

67   Marrack P, Kappler J: The staphylococcal enterotoxins and their relatives. Science 1990;248:705–711.

68   Fleischer B, Schrezenmeier H: T cell stimulation by staphylococcal enterotoxins: Clonally variable response and requirement for major histocompatibility complex class II molecules on accessory target cells. J Exp Med 1988;167:1697–1707.

69   Herrmann T, Maryansky JL, Romero P, Fleischer B, MacDonald HR: Activation of MHC class I-restricted CD8[+] CTL by microbial T cell mitogens: Dependence upon MHC class II expression of the target cells and V_β usage of the responder T cells. J Immunol 1990;144:1181–1186.

70 Sekali R-P, Croteau G, Bowman M, Scholl P, Burakoff Geha RS: The CD4 molecule
    is not always required for the T cell response to bacterial enterotoxins. J Exp Med
    1991;173:367–371.
71 Pullen A, Wade T, Marrack P, Kappler J: Identification of the region of the T cell
    receptor β chain that interacts with the self-superantigen Mls-1ᵃ. Cell 1990;61:
    1365–1374.
72 Cazenave PA, Marche P, Jouvin-Marche E, Voegtle D, Bonhomme F, Bandeira A,
    Coutinho A: $V_\beta 17$-gene polymorphism in wild-derived mouse strains: Two amino
    acid substitutions in the $V_\beta 17$ region alter drastically T cell receptor specificity. Cell
    1990;63:717–728.
73 Choi Y, Herman A, DiGusto D, Wade T, Marrack P, Kappler J: Regions of the
    variable region of the T-cell receptor β-chain that interact with S. aureus toxin
    superantigens. Nature 1990;346:471–473.
74 Pullen AM, Bill J, Kubo R, Marrack P, Kappler JW: Analysis of the interaction site
    for the self superantigen Mls-1ᵃ on T cell receptor V_β. J Exp Med 1991;173:1183–
    1192.
75 Fraser JD: High-affinity binding of staphylococcal enterotoxins A and B to HLA-DR.
    Nature 1989;339:221–223.
76 Mourad W, Scholl P, Diaz A, Geha R, Chatila T: The staphylococcal toxic shock
    syndrome toxin 1 triggers B cell proliferation and differentiation via major histo-
    compatibility complex-unrestricted cognate T/B cell interaction. J Exp Med 1989;
    170:2011–2022.
77 Herman A, Kappler JW, Marrack P, Pullen AM: Superantigens: Mechanism of T-cell
    stimulation and role in immune responses. Annu Rev Immunol 1991;9:745–772.
78 Janeway CAJ, Conrad PJ, Tite J, Jones B, Murphy DB: Efficiency of antigen
    presentation differs in mice differing at the Mls-locus. Nature 1983;306:80–82.
79 Hügin AW, Vacchio MS, Morse HC III: A virus-encoded 'superantigen' in a
    retrovirus-induced immunodeficiency syndrome of mice. Science 1991;252:424–
    427.

Dr. Hans Acha-Orbea, Ludwig Institute for Cancer Research,
CH–1066 Epalinges (Switzerland)

Fleischer B (ed): Biological Significance of Superantigens.
Chem Immunol. Basel, Karger, 1992, vol 55, pp 87–114

Mls Antigens: Immunity and Tolerance [1]

Susan R. Webb, Jane Hutchinson, Jonathan Sprent

Department of Immunology, IMM4A, The Scripps Research Institute,
La Jolla, Calif., USA

Introduction

Potent primary T-cell responses to what are now known to be murine
endogenous superantigens were first described nearly 20 years ago by Festenstein [1]. At that time, the nature of the stimulating antigens was unknown
and the antigens were termed 'minor lymphocyte-stimulating' (Mls) antigens. Through the efforts of a number of different laboratories, Mls antigens
were found to have highly unusual properties. Although the precursor
frequency of T cells responding to Mls antigens is as high or higher than that
for typical major histocompatibility complex (MHC) antigens [2–4], Mls
antigens do not appear to act as targets for graft-versus-host disease or to
elicit skin graft rejection [5, 6]. Mice cannot be primed to Mls antigens and,
as discussed later, injection of Mls^a-bearing cells into adult mice can induce a
long-lasting state of tolerance [7]. It was also noted that anti-Mls responses
display an unusual form of MHC restriction [reviewed in ref. 8]. Mls antigens
apparently have to be recognized in association with MHC (H-2) class II (Ia)
molecules, but either self or allo (foreign) Ia molecules can act as restricting
elements. Finally, it was found that clones of Mls -reactive T cells display an
apparently random distribution of H-2 alloreactivity [9]. This latter finding
provided perhaps the first clear indication that recognition of Mls antigens is
fundamentally different from recognition of conventional antigens.

During the past 4 years, several key breakthroughs have contributed to
our understanding of the immunobiology of Mls-specific responses. Many,

[1] Supported by grants CA41993, CA25803, CA38355 and AI21487 from the United
States Public Health Service. Publication No. 7179-IMM from the Scripps Research
Institute.

though not all, of the unusual features mentioned above can now be explained. In this article, we will review some of these recent studies with particular emphasis on the profound T-cell tolerance which occurs when either neonatal or adult mice are exposed to Mls antigens.

Nature of Mls and Related Endogenous Superantigens

Festenstein originally described Mls antigens as products of a single locus, the M locus (later termed the Mls locus), mapping to chromosome 1. Based on the ability of different strains to stimulate distinct patterns of T-cell responses in MHC-matched combinations, it was thought that there were 4 Mls alleles, Mls^a, Mls^b, Mls^c and Mls^d [10]. Mls^a- and Mls^d-encoded antigens were strongly stimulatory, whereas Mls^c-encoded products produced weak T-cell responses; strains typed as Mls^b were nonstimulatory. Subsequently, Mls^a and Mls^c genes were shown to map to different chromosomes, and strains typing as Mls^d were found to express not a different allele but rather both Mls^a and Mls^c antigens [reviewed in ref. 11]. To compound the confusion, Mls^c antigens mapped to multiple chromosomes, suggesting a much greater degree of genetic complexity than had been appreciated earlier [12, 13]. The terminology for Mls antigens is now rather confusing as some groups refer to Mls^a antigens as $Mls-1^a$ and to Mls^c antigens as $Mls-2^a$ or $Mls-3^a$ [12, 14, 15]. For simplicity we will use the original Mls designations; in view of the recent revelations on the nature of Mls antigens (see below), however, a radically new terminology for these antigens is in order.

A major breakthrough in our understanding of Mls antigens came with the generation of monoclonal antibodies to the T-cell receptor (TCR) Vβ chain. In mice expressing Mls^a antigen, it was discovered that nearly all T cells bearing Vβ6, Vβ8.1, Vβ9 and Vβ7 undergo clonal deletion in the thymus [16–20]. Likewise, T cells responding to Mls^a antigens in vitro and in vivo are markedly enriched for T cells expressing Vβ6$^+$ and Vβ8.1$^+$ TCR [16, 17, 21]. Since Vβ6$^+$ and Vβ8.1$^+$ cells collectively comprise about 15% of normal CD4$^+$ T cells, the high precursor frequency of T-cell responses to Mls^a antigens is readily explained. Anti-Mls^c responses are associated primarily with Vβ3$^+$ T cells, and Mls^c mice exhibit selective depletion of Vβ3$^+$ T cells [22–24]. In addition, it was found that mice showing joint expression of other related endogenous antigens plus I-E molecules are selectively depleted of Vβ11$^+$, Vβ5$^+$ and Vβ17$^+$ T cells [25–28]. Collectively, these findings indicate that, unlike TCR recognition of conventional peptide antigens which involve variable elements of both the α- and β-chain, the recognition of Mls and related antigens depends primarily on TCR Vβ expression. This dependence

Table 1. Vβ specificity of endogenous Vβse

Retrovirus	Vβ specificity
mtv-1	3
mtv-2	14
mtv-3	3
mtv-6	3, 5
mtv-7	6, 8.1, 7, 9
mtv-8	5(?), 11
mtv-9	5, 11
mtv-11	11
mtv-13	3

Data are taken from references 28, 31–34, 39 and 40.

on only one chain of the TCR implies that the interaction of the TCR with Mls and Mls-like antigens – henceforth termed Vβ-selecting elements (Vβse) [8] – is quite different from TCR interactions with peptide antigens. In this respect, it is now clear that the regions critical for TCR interaction with Vβse lie outside of the predicted TCR combining site [29, 30] (see below).

The second major breakthrough on the nature of Vβse came with the discovery that these antigens are encoded by endogenous mouse mammary tumor virus (MMTV) sequences (table 1). Following up the earlier suggestion that the co-ligand for Vβ5$^+$ T cells maps to an endogenous *mtv* gene on chromosome 12 [28], Frankel et al. [31] carried out a thorough genetic analysis which showed perfect concordance between the gene for the Mls^a antigen and the *mtv-7* provirus sequences found on chromosome 1. These same workers found that the complexity of Mls^c antigens (see above) could be accounted for by expression of any of 3 MMTVs, *mtv-1, mtv-6* and *mtv-13*. Parallel studies of Dyson et al. [32], Woodland et al. [33] and Acha-Orbea et al. [34] showed association of other Vβse with various combinations of other endogenous provirus sequences, and Marrack et al. [35] demonstrated that infectious MMTVs (transmitted in milk) express Vβse recognized by Vβ14$^+$ T cells. It should be mentioned that expression of Vβse may not be restricted to MMTV as recent studies implicate Vβse in the marked immunodeficiency associated with the murine leukemia virus mouse AIDS model [36] as well as human HIV infection [37].

A particularly interesting feature of MMTV-encoded Vβse is that expression depends on the open reading frame (ORF) located in the 3′ long terminal repeat [34, 38, 39]. Thus, transfection of this gene is sufficient to confer the ability to stimulate T-cell hybridomas expressing the appropriate $V\beta^+$ TCR. Comparison of the predicted amino acid sequences of ORFs from different MMTVs suggests a high degree of overall similarity. Nonetheless, significant variation in sequence is seen in the C-terminal region, suggesting that this region may control $V\beta$ selectivity [reviewed in ref. 40]. Previous studies have provided evidence suggesting that MMTV ORF gene products encode transcription factors, implying that ORF genes do not encode $V\beta$se but merely control the expression of these antigens [41, 42]. Though surface expression of ORF products has not been formally demonstrated, the notion that ORF products act simply as transcription factors is difficult to reconcile with $V\beta$ specificity.

Bacterial Superantigens

Though a thorough review of the literature on bacterial toxins is outside the scope of this article, it is important to mention that, like retroviruses, many bacteria express $V\beta$se that induce profound immunological effects. As for cell-bound $V\beta$se, these soluble toxins are strongly immunogenic and selectively stimulate T cells bearing particular $V\beta$ gene segments; different toxins display specificity for distinct but overlapping groups of $V\beta$s [43–48].

Like the retroviral superantigens, bacterial toxins require class II molecules for presentation [43–46, 48] and, when injected in vivo, induce tolerance of developing and mature T cells (see below) [49–52]. Although both classes of superantigens have $V\beta$ specificity, it is of interest that bacterial toxins bear no structural resemblance to endogenous retroviral ORF gene products. At this point, there are few clues as to how and why viruses and bacteria have both evolved distinct $V\beta$-selecting proteins. In the case of bacterial superantigens, the hyperacute response (e.g. shock) evoked by these antigens might facilitate survival of the bacteria by interfering with conventional antigen-specific responses, for example, by inactivating a large proportion of T cells or causing the release of toxic levels of T-cell-derived cytokines. An interesting notion suggested by Janeway [8] is that bacterial toxins may act by appropriating the endogenous MMTVs themselves. The particular $V\beta$ specificity of individual bacterial toxins, however, does not correlate with the $V\beta$ specificities of any of the known endogenous MMTVs, which may suggest a rather more complicated explanation.

It is important to note that the potent physiological effects of bacterial toxins may not be solely a consequence of T-cell activation. These proteins also stimulate marked production of non-T-cell-derived cytokines such as interleukin-1 (IL-1) and tumor necrosis factor [53–56]. At least in the case of macrophages, activation of antigen-presenting cells (APCs) presumably occurs as a consequence of toxin binding to class II antigens, although this point has not been formally documented; activation of B cells, on the other hand, appears to require T-B-cell interaction [57]. There is no evidence that endogenous MMTV expression leads to APC activation. On the contrary, MMTV expression appears to be up-regulated on APCs as the result of APC activation by external stimuli. Thus, stimulation of B cell tumors or splenic B cells with lipopolysaccharide or certain cytokines (e.g. IL-4) leads to the transcriptional activation of MMTV [33, 58, 59]. Although MMTV expression is low on resting B cells, it is clear that Mls^a *(mtv-7)* expression on resting B cells is sufficient to cause T-cell activation. This interaction also results in backstimulation of the B cells. Thus, T-cell recognition of Mls antigens on resting B cells is able to stimulate the B cells to undergo marked proliferation. Based on experiments with a dual reactive T-cell clone specific for both Mls^a and H-2^p alloantigens, we have found that recognition of either Mls^a or H-2^p antigens on B cells leads to B cell proliferation [Webb, unpubl. data]. Interestingly, however, only the H-2 stimulus drives the B cells to differentiate to cytoplasmic immunoglobulin-positive cells. This finding raises the intriguing possibility that T-cell interaction with endogenous MMTV gene products delivers a unique set of signals to B cells resulting in activation without the initiation of immunoglobulin gene transcription. Alternatively, the data may reflect distinct signals resulting from T-B-cell interactions of differing avidities.

Antigen-Presenting Cells

Dendritic cells [60] are by far the most potent APCs for T-cell recognition of typical class-II-restricted peptides. Macrophages and activated B cells can also present peptides [reviewed in ref. 61], especially to previously activated T cells. In contrast, the ability to stimulate Mls-reactive T cells appears to be largely limited to B cells [62–64]. As mentioned above, resting B cells, which are poor APCs for conventional antigens, are able to generate Mls-specific proliferative responses by resting T cells [65]. The explanation for this difference in the APC requirements for Vβse versus 'conventional' antigens is not immediately apparent, but could reflect that Mls-specific T-cell responses have different co-stimulatory requirements from responses to conventional antigens. In this regard, there is evidence that the failure of resting B cells to stimulate responses to H-2 and other conventional antigens can be attributed at least in part to lack of expression of B7/BB1, the B cell ligand for CD28 (an accessory molecule on T cells with signal transduction function) [reviewed in ref. 66]. Whether B7-CD28 interaction is required for anti-Mls responses is unknown.

In the case of other cell types, the issue of whether dendritic cells and macrophages can stimulate anti-Mls[a] responses is controversial [64, 67–69]. We ourselves have found no evidence that highly purified macrophages or dendritic cells are capable of eliciting anti-Mls[a] responses in vitro by T-cell clones or T hybridomas [64]. Although lack of antigen expression on these APCs would seem the most likely explanation for this finding, it should be noted that, at present, there is no direct evidence on the expression of MMTV transcripts by macrophages or dendritic cells. Nevertheless, the limited Mls stimulation by these professional APCs suggests that Mls[a] antigens are not subject to typical antigen processing and cannot be readily absorbed from other cells in sufficient quantities to stimulate T hybridomas. Some groups contest this view and argue that experiments with bone marrow chimeras provide strong evidence that Mls antigens can be transferred [24, 70, 71]. Our own studies with chimeras, however, have shown no evidence that Mls antigens can move from one cell to another [64]. Moreover, in the case of bacterial toxins there is no convincing evidence that these antigens require antigen processing. Thus, fixed APCs are able to present these toxins as well as unfixed cells [43, 44]; furthermore, digestion of bacterial superantigens by proteolysis tends to reduce rather than amplify their immunogenicity [72].

At present, the precise receptor-ligand interactions involved in T-cell recognition of Vβse are still poorly understood, and most of the available evidence has come from studies with bacterial toxins. With these antigens there is evidence that the contact residues on class II molecules are not the same as for typical peptide-MHC interactions. Whereas typical peptides bind in the MHC 'groove', the data of Dellabona et al. [73] support the idea that bacterial enterotoxins bind to the lateral face of class II molecules; Janeway et al. [44] have postulated that enterotoxins also make contact with the lateral face of the TCR Vβ. Such joint binding is presumed to allow these Vβse proteins to act as a sort of TCR-MHC clamp, thus causing T cells and APCs to form conjugates. There is also evidence that Mls antigens may cause similar bridging of TCR and class II molecules [29, 30]. Interestingly, however, analysis of the predicted amino acid sequence of several of the MMTV ORFs suggests a structure consistent with a type II membrane protein [N. Gascoigne, pers. communication]. This suggests that Mls antigens might be integral cell-bound proteins rather than soluble peptides.

As for other antigens, the trimolecular interaction of Vβse, TCR and H-2 molecules is presumed to be complemented by additional interactions between accessory molecules on T cells and APCs (exemplified by LFA-1-ICAM, CD28-B7 interaction). This raises the question whether responses to Vβse involve both CD4$^+$ and CD8$^+$ T cells. CD4 and CD8 molecules bind to monomorphic sites on H-2 class II and class I molecules, respectively, thereby strengthening T-APC interaction as well as providing critical signal transduction via their intracytoplasmic association with the tyrosine kinase p56^{lck} [74]. Since Vβse associate preferentially with class II molecules, one would expect Vβse responses to be restricted to CD4$^+$ cells. This is only partly true. When purified CD4$^+$ and CD8$^+$ cells are tested separately, CD4$^+$ cells give very strong proliferative responses to Mls^a antigens, whereas CD8$^+$ cells give virtually no response. When mixed with CD4$^+$ cells or soluble lympho-kines, however, CD8$^+$ cells do give significant responses, both in vitro and in vivo [21, 75]. The simplest explanation for this finding is that CD8$^+$ cells recognize Mls^a/Ia complexes with low avidity. This weak interaction induces only partial triggering and the cells fail to proliferate unless supplemented with exogenous lymphokines.

Although TCR specificity for Vβse is largely a reflection of Vβ specificity, it is quite likely that other TCR elements (e.g. Vα) influence the overall avidity of TCR-Vβse interaction. For this reason, T cells expressing a given Vβ TCR probably recognize Vβse with a spectrum of different avidities. This issue will be discussed later in the context of T-cell tolerance of Vβse.

As a final point, it should be emphasized that the strong primary response of T cells to Mls antigens does not apply to other endogenous Vβse. Thus, the MMTV products recognized by Vβ5$^+$ and Vβ11$^+$ T cells induce only negligible responses in vitro – which accounts for why these antigens were not discovered earlier. Nevertheless, under in vivo conditions, the Vβse recognized by Vβ5$^+$ and Vβ11$^+$ cells do elicit appreciable proliferative responses [76]. Why these antigens are poorly immunogenic in vitro is unknown. The most obvious possibility is that such antigens lead to only very low-avidity T-APC interactions.

Intrathymic Tolerance

T-cell differentiation in the thymus involves a complex process of positive and negative selection [61]. Positive selection occurs at an early stage of differentiation and allows the selective survival of T cells bearing

TCR which interact with the MHC molecules expressed on cortical epithelial cells; positive selection ensures then that the T cells leaving the thymus will recognize antigen in association with 'self'-MHC (display self-MHC restriction). Negative selection removes T cells which react with self-MHC antigens too strongly, thus protecting the host from potential autoaggressive T cells.

Much of what has been learned about how thymic selection regulates repertoire development has come from the study of Vβse. For example, one of the first formal demonstrations that negative selection occurs in the thymus itself and involves clonal deletion was the observation that thymocytes bearing high levels of Vβ17+ TCR are selectively removed in I-E+ mouse strains [25]. Immature thymocytes expressing low levels of Vβ17+ TCR are not deleted, which suggests that clonal deletion occurs at a relatively late stage of development. Similar studies showed that other Vβse cause clonal deletion in the thymus. Thus, Vβ6+ and Vβ8.1+ cells are selectively deleted in Mlsᵃ-bearing mice, whereas Vβ3+ cells are absent from Mlsᶜ mice [16, 17, 22–24]. Like Vβ17+ T cells, Vβ5+ and Vβ11+ cells are eliminated in I-E+ mouse strains (providing these strains express the appropriate endogenous MMTV) [26–28]. It is of interest that in all of these cases Vβ deletion applies to both CD4+ and CD8+ T cells. Since the response of mature T cells to Vβse is largely restricted to CD4+ T cells, intrathymic clonal deletion is presumed to occur at the level of CD4+CD8+ thymocytes. In support of this notion, Fowlkes et al. [77] found that neonatal injection of anti-CD4 antibody allowed the development of CD8+ T cells bearing Vβ17+ TCR. Similarly, MacDonald et al. [78] showed that anti-CD4 antibody blocked the negative selection of CD8+ Vβ6+ T cells in Mlsᵃ mice. It is notable, however, that only a portion of CD4+CD8+ thymocytes appear to be susceptible to deletion by most Vβse [79]. This finding contrasts with the observation that T-cell tolerance to conventional (non-Vβse) antigens can occur very early in ontogeny [80]. This is clearly illustrated in transgenic mice expressing a Vβ8.1+ TCR specific for both lymphocyte choriomeningitis virus and Mlsᵃ [81]. For this transgenic line, intrathymic expression of virus causes marked deletion of CD4+CD8+ immature thymocytes, whereas Mlsᵃ expression deletes primarily mature thymocytes. Why Vβse fail to delete most CD4+CD8+ thymocytes is still unclear. One possibility is that the avidity of TCR-antigen interactions is lower for Vβse than for other antigens. Alternatively, the expression of Vβse in the thymus might be largely restricted to the medulla, i.e. the site of mature T cells (see below). Both of these factors may play a role in determining the precise stage of negative selection for individual T cells.

Which Cell Types in the Thymus Control Negative Selection?

The importance of bone-marrow-derived APCs, especially macrophages and dendritic cells, in the induction of intrathymic tolerance to conventional MHC-restricted antigens is well known. Thus, T cells developing in chimeric thymuses where antigen expression is limited to the epithelial cells tend to display only limited tolerance [reviewed in ref. 82]. By contrast, full tolerance occurs when antigen is displayed on APCs, especially on dendritic cells. Although thymic epithelial cells (TECs) are clearly less tolerogenic than APCs, TECs do play a significant role in tolerance induction, especially for CD4+ T cells. Under conditions where antigen is displayed only on TECs and not on APCs, the CD4+ cells which escape tolerance respond poorly in mixed lymphocyte reactions, are unusually sensitive to the inhibitory effects of anti-Ia antibodies and fail to reject skin grafts on adoptive transfer [82]. These properties suggest that the cells which evade tolerance induction are of low affinity.

Which cell types in the thymus control tolerance induction to Mls antigens is complex. For Mls^a antigens, TECs seem to play little if any role in tolerance induction. This is apparent from the finding that T cells developing in athymic (nude) Mls^b mice given APC-depleted (deoxyguanosine-treated) thymuses from Mls^a donors show no Mls^a tolerance and no deletion of Vβ6+ cells; normal thymus grafts, by contrast, induce full tolerance and Vβ6+ deletion [83]. Since deoxyguanosine treatment of thymuses removes T and B cells in addition to APCs, any of these cell types might control or contribute to Mls^a tolerance induction.

Since the capacity to stimulate Mls^a-reactive mature T cells in vitro is largely limited to B cells (see above), it would seem highly likely that B cells play an important role in Mls^a tolerance induction. In support of this idea, injecting neonatal Mls^b mice with highly purified B cells from adult Mls^a mice induces strong tolerance to Mls^a antigens and elimination of Vβ6+ cells [84]. With regard to other cell types, there is conflicting evidence on whether dendritic cells contribute to Mls^a tolerance. One group has reported that dendritic cells are effective [68], whereas another group finds that dendritic cells are only effective when supplemented with B cells [85]. T cells would seem unlikely candidates for inducing Mls^a tolerance because these cells are generally assumed to be Ia$^-$ (in mice) and are poor stimulators of anti-Mls^a responses in vitro. Surprisingly, however, T cells – especially CD8+ cells – are strongly tolerogenic [84]. This was demonstrated by injecting neonatal Mls^b mice with highly purified populations of CD4+ or CD8+ cells from Mls^a mice. Intravenous injection of as few as 2×10^4 CD8+ cells at birth leads to near-

Table 2. Intrathymic T-cell chimerism in Mls^b mice injected at birth with Thy-1-marked Mls^a CD8$^+$ cells given intravenously

Neonatal mice tested	Number of cells injected	Percent of T cells staining for Thy-1.1		
		Jlld$^-$ thymus		lymph nodes 2 weeks
		1 week	2 weeks	
B10.BR (Thy-1.2$^+$) control	–	0.1	0.3	0.1
B10.BR injected with Mls^a CD8$^+$ T cells from AKR/J (Thy-1.1$^+$) mice	10×10^6 2×10^6	2.3 NT	1.5 0.7	24.0 7.2

As described elsewhere [84], neonatal B10.BR mice were injected intravenously within 24 h of birth with 10×10^6 or 2×10^6 AKR/J (H-2^k, Mls^a, Thy-1.1) CD8$^+$ T cells. Thymus cell suspensions were depleted of immature T cells by treatment with Jlld antibody plus complement. The remaining mature thymocytes were stained with biotin-conjugated anti-Thy-1.1 monoclonal antibody (19E12) plus FITC-streptavidin and analyzed by flow cytometry. [Previously unpublished data of S. Webb.]

complete Mls^a tolerance and strong Vβ6$^+$ T-cell deletion. Cell for cell, CD8$^+$ cells are 50- to 100-fold more tolerogenic than CD4$^+$ cells or B cells. Tolerance induced in this system is evident in both thymus and lymph nodes, which indicates that tolerance is induced intrathymically.

Evidence that intravenously injected CD8$^+$ cells reach the thymus in intact form (rather than as cell fragments) has come from studies with Mls^b neonates injected with purified T cells taken from Thy-1-marked Mls^a mice (AKR/J → B10.BR) [C. Surh, Webb, Sprent, unpubl. data]. As shown in table 2, significant numbers of T cells carrying the donor Thy-1 marker are detectable in the mature (Jlld$^-$) component of the host thymus at 1–2 weeks after injection. Homing of donor T cells to the host thymus is detectable within 1 day of injection and applies to resting T cells as well as activated cells (which contrasts to the adult thymus where thymus homing is restricted to activated T cells) [86]. In histological sections, the donor T cells are largely restricted to the medulla rather than the cortex [Surh, Webb, Sprent, unpubl. data]. The selective homing of the immigrant T cells to the medulla correlates with the finding that the deletion of Vβ6$^+$ cells in normal Mls^a mice occurs relatively late in differentiation (see above).

The above data indicate that, in adoptive transfer systems, intrathymic migration of lymphoid cells (CD8$^+$ cells and B cells) from the peripheral lymphoid organs can cause strong intrathymic tolerance and deletion. Are these data relevant to tolerance induction occurring in situ, i.e. to self Mls^a in the normal Mls^a thymus? In speculating on this important issue, it is of interest that intrathymic deletion of Vβ6$^+$ cells in Mls^a mice is minimal at birth and does not become conspicuous until 1–2 weeks of age [87]. Four possibilities could account for this late deletion of Vβ6$^+$ cells during ontogeny: (1) impaired Mls reactivity of neonatal T cells; (2) paucity of the appropriate APCs in the neonatal thymus; (3) delayed tissue expression of Mls antigens, or (4) some unknown developmental stromal cell defect perhaps related to important cytokines.

With regard to the first possibility, it is known that, unlike adult T cells, neonatal T cells lack significant TCR N-region diversity [88]. Whether N-region diversity (which fine-tunes T-cell specificity) affects Vβse reactivity is unclear. Neonatal Mls^b thymocytes give quite strong proliferative responses to Mls^a antigens in vitro, but this finding does not exclude the possibility that neonatal T cells have a somewhat lower avidity for Mls^a antigens than adult T cells.

The possibility that the neonatal thymus lacks appropriate APCs for Mls deletion might seem unlikely since all of the cell types found in the adult thymus appear to be represented in the neonatal thymus. Nevertheless, it has been found that selectively depleting Mls^a mice of B cells by injecting anti-μ antibody from birth impairs the deletion of Vβ6$^+$ cells, especially when anti-μ treatment is begun before birth (by treating the mothers with anti-μ antibody, thereby creating second-generation μ-suppressed mice) [Webb, unpubl. data]. This finding would seem to suggest that B cells play an obligatory role in causing complete tolerance induction to Mls^a antigens – a clear contrast to the system described above where injecting neonatal Mls^b mice with purified Mls^a CD8$^+$ cells was sufficient to cause near-complete deletion of Vβ6$^+$ cells. Although this apparent contradiction has yet to be resolved, it is worth pointing out that μ suppression of mice markedly impairs T-cell activation (which has been taken as evidence that B cells play an important APC function in vivo) [61]. Most of the CD8$^+$ T cells in μ-suppressed mice might thus be resting cells and, therefore, be poorly immunogenic. As discussed below, Mls^a expression is much lower on resting CD8$^+$ cells than on activated cells.

With regard to the third possibility, the notion that the impaired deletion of Vβ6$^+$ cells in Mls^a neonates reflects delayed antigen expression in the thymus is difficult to assess. One approach to this question is to examine the tempo of Vβ6$^+$ cell deletion in neonatal Mls^b mice injected with CD8$^+$ cells

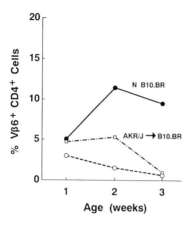

Fig. 1. Delayed intrathymic deletion of Vβ6+ cells in Mlsᵇ mice injected with Mlsᵃ CD8+ cells at birth. B10.BR (Mlsᵇ) mice were injected intravenously within 24 h of birth with a dose of 2×10^5 (□) or 2×10^6 (○) purified CD8+ cells from Mlsᵃ (AKR/J) mice. At the indicated times, thymocytes were separated into mature (Jlld⁻) cells with antibody and complement (see table 2) and double-stained for Vβ6 versus CD4 expression [84]; Jlld⁻ thymocytes from (N) uninjected age-matched B10.BR mice were used as a control (●). The data are shown in terms of the percentage of CD4+ cells that express Vβ6+ TCR. All groups exhibited comparable expression of Vβ8.2 (not shown). [Previously unpublished data of S. Webb.]

from adult Mlsᵃ mice. The striking finding here is that Vβ6+ cell deletion is still quite limited at 1 week after injection (fig. 1). Thus, whereas injection of as few as 1×10^5 cells causes complete deletion at 3 weeks after injection, even 10-fold higher doses of CD8+ cells cause only partial deletion at 1 week. Yet, as discussed earlier, the injected cells reach the host thymus within 24 h. Why then is deletion of Vβ6+ cells inefficient? It is conceivable that tolerogenicity of CD8+ cells depends on activation-induced up-regulation of Mlsᵃ *(mtv-7)* expression. In this respect, it is known that *mtv-7* transcript expression is very low in resting CD8+ T cells but increases considerably following T cell activation [B. Huber, pers. communication]. Thus, the tolerogenicity of normal CD8+ cells (most of which are resting cells) might require that the CD8+ cells reaching the thymus are in an activated state. Generating such activated T cells in vivo might take several days. The important question, therefore, is whether injecting Mlsᵇ mice with *preactivated* Mlsᵃ CD8+ cells induces a more rapid onset of Vβ6+ cell deletion in the thymus. This question is currently under study.

The final possibility is that defective tolerance to Mls^a antigens in the neonatal period reflects an intrinsic and general inability of the neonatal thymus to cause deletion, e.g. because of a failure to synthesize certain requisite cytokines and/or accessory molecules. The critical question here is whether the neonatal thymus is capable of deleting T cells specific for conventional (non-Vβse) antigens. Surprisingly, very little direct evidence is available on this question.

At present, none of the 4 possibilities discussed above can be ruled out. The lack of Vβ deletion in the neonatal thymus thus remains a paradox. With regard to postnatal tolerance, the evidence that T cells (CD8$^+$ cells) and B cells induce potent Mls^a tolerance in adoptive transfer models suggests that these cells probably play a crucial role in tolerance induction, at least for Mls^a antigens. Although other cell types in the thymus such as macrophages, dendritic cells and epithelial cells seem to make only a minor contribution to Mls^a tolerance, the possibility that these cells are involved in inducing tolerance to other Vβse cannot be excluded.

Extrathymic Tolerance

As discussed above, clonal deletion in the thymus is an efficient method for purging the mature T-cell repertoire of potentially autoreactive T cells. Nevertheless, it could be argued that clonal deletion in the thymus can never be complete and that some autoreactive cells will inevitably reach the periphery. In discussing this issue, it is essential to make a distinction between common blood-borne self-antigens and other tissue-specific self-antigens which, under normal circumstances, fail to make contact with resting T cells. In the case of soluble or cell-bound antigens in the blood, these antigens are known to have access to the thymic medulla and thus make close contact with immature T cells. Such contact is presumed to lead to clonal deletion in situ, the efficiency of deletion reflecting the amount of antigen reaching the thymus. This line of reasoning predicts that antigens entering the thymus in only low concentrations cause only partial deletion. In support of this possibility, injection of small numbers of Mls^a-bearing cells (e.g. 10^4) into Mls^b neonates leads to partial intrathymic deletion of Vβ6$^+$ cells, whereas injection of large numbers of cells (e.g. 10^5) leads to complete deletion [Webb, unpubl. data]. With limiting doses of antigen, some of the Vβ6$^+$ cells that evade deletion reach the periphery. These cells are probably low-avidity cells, high-avidity cells having been deleted. The escape of residual low-avidity cells from the thymus is unlikely to be dangerous, because the concentration of antigen encountered in the periphery is presumably below the threshold required for triggering these cells; these cells would only be reactive to very high concentrations of antigen. In the case of tissue-specific antigens, the concentration of these antigens in the thymus is probably too low to cause more than minimal intrathymic deletion. T cells specific for these antigens leave the thymus but rarely cause autoimmunity, presumably because tissue-specific self-antigens normally remain sequestered and fail to enter the lymphoid tissues (the sites of T-cell induction).

The notion that T-cell tolerance to soluble self-antigens is largely a reflection of clonal deletion in the thymus hinges on the assumption that all of these various self-antigens reach the thymus in sufficient quantities to tolerize early T cells. Though the thymus is highly permeable to soluble antigens (and APCs), the entry of such antigens into the thymus might sometimes be inefficient, with the result that the concentration of self-antigen would be appreciably lower in the thymus than in the peripheral lymphoid tissues. There would then be the danger that the nondeleted T cells leaving the thymus would encounter sufficiently high concentrations of antigen in the spleen or lymph nodes to cause T-cell triggering and thereby break self-tolerance. Because of this potential danger, maintenance of self-tolerance may require backup mechanisms in the periphery to prevent autoaggression. This is incontestable in the case of Mls and related endogenous superantigens since clonal deletion operates poorly in the neonatal period (see above). How these potentially autoreactive T cells are kept under control is discussed below.

Suppression

Beginning with the pioneering studies of Gershon et al. [89], the notion that peripheral self-tolerance might be maintained by a distinct subset of T cells with suppressor function has generated considerable attention [90, 91]. However, despite intensive investigation, the role of antigen-specific suppressor cells in self-nonself discrimination remains controversial and confusing. Indeed, there is little unequivocal evidence that suppressor cells exist as a separate lineage. Recent studies raise the possibility that much of the evidence on suppression may be attributed to typical $\alpha\beta$ TCR$^+$ cells inhibiting the function of other immune cells (e.g. through cytotoxic activity) and to the production of inhibitory cytokines by subsets of helper T cells [92–94]. Reviewing the role of suppressor cells and factors is beyond the scope of this article. Since we have found no evidence for suppression in Mls tolerance, the remainder of this section is devoted to a discussion of other possible mechanisms of tolerance.

Anergy

The studies of Jenkins, Schwartz and others [95, 96] have provided convincing evidence that mature T cells can be rendered tolerant through induction of anergy. These investigators have shown that T-cell clones exposed to fixed antigen-pulsed APCs can be rendered unresponsive (anergic) to subsequent challenge with normal (unfixed) antigen-pulsed APCs. The addition of IL-2 can partially overcome this unresponsiveness, but neither IL-2 nor a variety of other cytokines and factors prevent the induction of anergy. However, addition of third-party APCs, i.e. cells which cannot present the antigen in question does prevent the induction of anergy. This finding suggests that the missing 'second signal' required for normal T-cell induction can be provided by bystander cells. In this respect, it has long been argued that T-cell stimulation by APCs requires that, in addition to antigen (signal 1), the APCs must display co-stimulatory function (signal 2); only 'professional' APCs such as dendritic cells, macrophages and activated B cells are thought to express this function [reviewed in ref. 96]. In the absence of co-stimulation, T cells enter an anergic state. Although the nature of the co-stimulatory signal of APCs is not definitively resolved, recent evidence suggests that interaction of the T cell accessory molecule CD28 with its ligand BB1/B7 on APCs may be critical for productive T-cell activation [66, 97].

Evidence that anergy operates in vivo has come from studies using transgenic mouse models. Thus, in transgenic lines in which foreign I-E antigens are expressed selectively in sites such as the stromal cells of the pancreas (i.e. on nonprofessional APCs), tolerance to I-E molecules is observed [98]. T cells expressing Vβ5$^+$ or Vβ17$^+$ TCR are not deleted in these mice. Nonetheless, these T cells are markedly unresponsive to anti-TCR stimulation, indicative of anergy. Evidence that such anergy induction can apply to antigens other than Vβse has come from comparable studies in transgenic mice expressing foreign class I (K^b) antigens in the pancreas. When bred with K^b-specific TCR transgenic mice, the resulting double-transgenic mice do not show T-cell deletion [99]. However, the nondeleted T cells fail to generate K^b-specific cytotoxic T-lymphocyte responses. These and other studies [100] support the notion that maintenance of self-tolerance of mature T cells can involve induction of an anergic state. Drawing firm general conclusions from these studies, however, is difficult, because other groups have failed to see tolerance in other comparable lines of transgenic mice [reviewed in ref. 101]. Moreover, there is the conceptual difficulty of explaining how selective expression of antigen in the pancreas induces tolerance in all T cells throughout the body, including the thymus in some studies; to account for such pervasive tolerance, one is forced to postulate that all T cells, including mature thymocytes, migrate through the pancreas. This seems rather unlikely. Other evidence for anergy in vivo has come from studies with transgenic mice expressing the Mls^a-reactive Vβ8.1$^+$ TCR [102]. These mice show incomplete deletion of Vβ8.1$^+$ cells, and the nondeleted cells are tolerant to Mls^a antigens and fail to respond normally to anti-TCR stimulation. The nature of the APCs inducing anergy in this model is unknown.

These and other studies have led to the notion that, whereas clonal deletion is the major mechanism for intrathymic tolerance, clonal anergy controls tolerance outside of the thymus. Though compartmentalization of these different mechanisms is attractive, there are conceptual problems with the notion that anergy plays a decisive role in self-nonself discrimination in the extrathymic environment. As discussed above, anergy induction is presumed to depend on T-cell contact with cells lacking requisite second signals, e.g. stromal cells of the pancreas. But one would expect many self-antigens to be expressed on professional APCs and thereby be presented in immunogenic rather than tolerogenic form. Second, anergy models based on inappropriate presentation of antigen require that T-cell interactions with self-antigens occur in a microenvironment devoid of professional APCs; otherwise, bystander co-stimulation would occur and lead to immunogenicity. Third, since IL-2 allows anergic cells to proliferate, this type of tolerance would be tenuous at best and likely to break down in the face of T-cell stimulation and IL-2 production occurring in response to infectious pathogens. Fourth, unless anergic cells are eventually destroyed, these cells would be expected to accumulate and thereby clutter the immune system with useless cells.

Deletion

Until recently, clonal deletion of T cells was thought to be restricted to the thymus. As discussed below, however, clonal deletion can also operate outside of the thymus. Our own interest in this idea stemmed from the observation of Lilliehöök et al. [7] that injection of adult Mls^b mice with H-2-matched Mls^a cells led to a profound and prolonged state of Mls^a-specific tolerance; the mechanism of tolerance was not defined. Subsequently, Rammensee et al. [103] reported that injection of Mls[a] spleen cells into Mls[b] mice failed to induce Vβ6[+] cell deletion. Since the mice were fully tolerant to Mls[a] antigens, it was concluded that tolerance reflected anergy.

We have recently repeated these experiments using thymectomized recipients [104]; in this respect, the use of adult thymectomized hosts avoids the problem of de novo production of T cells by the thymus during the course of the experiment. In confirmation of the studies of Lilliehöök et al. [7] and Rammensee et al. [103], we found that injection of Mls[b] mice with MHC-matched Mls[a] lymphoid cells (AKR/J → B10.BR) leads to profound Mls-specific functional tolerance as measured in primary mixed lymphocyte reactions. However, in contrast to the studies of Rammensee et al. [103], by 2–3 weeks after transfer we observed a marked reduction (from 50 to 80%) in the number of host CD4 cells expressing Vβ6 relative to the preinjection levels; the disappearance of the Vβ6[+] cells is limited to CD4[+] cells. These data indicate that, under certain circumstances, extrathymic contact of mature T cells with antigen can lead to clonal elimination of the reactive cells. Extrathymic deletion of mature T cells has also been seen in mice injected with staphylococcal enterotoxin B (SEB), a superantigen which is strongly immunogenic for Vβ8[+] T cells in vitro [52]. By 1 week after injection, SEB-injected mice show a selective reduction in the number of Vβ8[+] T cells; again, the elimination of these cells is restricted to CD4[+] cells. In other situations, extrathymic deletion of mature T cells can affect CD8[+] cells. Thus, male-antigen (H-Y)-specific CD8[+] cells from TCR-transgenic mice undergo rapid elimination when transferred to male (but not female) hosts [105]. Extra-thymic deletion can also apply to T cells escaping intrathymic tolerance induction in the neonatal period. It has been shown that injecting Mls[a] mice from birth with anti-I-E antibody allows Vβ6[+] cells to reach the peripheral lymphoid tissues [106]. Significantly, terminating the injections of antibody causes these Vβ6[+] cells to disappear.

Except in this last model, in each of the situations described above the elimination of T cells in the postthymic environment affects only one of the two major T-cell subsets. For example, in Mls[b] mice injected with Mls[a] lymphoid cells, the elimination of host Vβ6[+] cells is much more prominent for CD4[+] cells than CD8[+] cells. This finding correlates with the observation that CD4[+] cells give much stronger responses to Mls[a] antigens in vitro than CD8[+] cells. The susceptibility of T cells to extrathymic deletion thus seems to be related in some way to the level of antigen reactivity of the participating cells, cells with strong reactivity to antigen being more susceptible to deletion than cells with weaker reactivity. Such a correlation implies that clonal

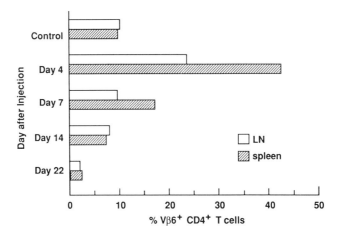

Fig. 2. Clonal elimination of Vβ6⁺ T cells in adult Mlsᵇ mice injected with Mlsᵃ CD8⁺ cells. Adult-thymectomized B10.BR (Mlsᵇ) mice were injected intravenously with a dose of 10^7 purified CD8⁺ cells from Mlsᵃ (AKR/J) mice. At the times indicated, spleen and lymph node (LN) cells were depleted of donor (Thy-1.1⁺) T cells and analyzed for Vβ6 expression at the level of host CD4⁺ cells using two-color fluorescence and flow cytometry. Vβ8 expression remained constant throughout the experiment (not shown). [Data adopted from ref. 104.]

elimination is preceded by an immune response. This is indeed the case. Thus, except in the neonatal tolerance model, in each of the situations discussed above the elimination of T cells is preceded by a phase of T-cell proliferation. For example, when Mlsᵃ lymphoid cells are transferred to Mlsᵇ mice of the H-2ᵏ haplotype, host Vβ6⁺ cells initially undergo marked expansion with the result that, at 2–4 days after transfer, Vβ6⁺ cells account for up to 40% of total CD4⁺ cells in spleen [104]; these cells then gradually disappear to reach levels of 2–4% by 3 weeks after transfer (fig. 2).

The key conclusion emerging from the above data is that apparently normal immune responses in vivo can eventually culminate in extensive elimination of the responding T cells. What happens to the stimulated cells? There are two broad possibilities (which are not mutually exclusive).

The first possibility is that the disappearance of T cells from the lymphoid tissues (spleen and lymph nodes) reflects T-cell migration from these tissues to other sites in the body, e.g. the intestines. Populations of activated T cells are known to home in large numbers to the gut after intravenous injection and some of these cells appear to be excreted into the gut lumen [107]; whether some of the T cells reaching the gut eventually return to the spleen and lymph nodes is unclear.

The second possibility is that the elimination of T cells is a reflection of overstimulation leading to cell death. In support of this idea, it has been found that many of the activated $V\beta 8^+$ T cells in the spleens of mice given SEB 4 days before are in the process of dying [52]. Cell death reflects a metabolically active process of 'apoptosis' in which endonucleases in the cells are activated to digest DNA. Although apoptosis of T cells is now well documented, the precise triggering events which initiate this process are poorly defined. We speculate that strong triggering may cause the elevation of certain intracellular second messengers which directly or indirectly initiate apoptosis. In this respect, we have found that exposing T cells to limiting doses of SEB in vitro causes strong proliferation but little detectable phosphatidylinositol (PI) turnover [A. O'Rourke and S. Webb, unpubl. data]. Conversely, stimulating T cells with larger doses of SEB leads to impaired proliferative responses accompanied by easily detectable activation of the PI pathway. This finding raises the possibility that high levels of PI metabolites (or the subsequent rise in intracellular $[Ca^{2+}]$) may be inversely correlated with immune responses and, in fact, may actually be toxic to the cell. Direct evidence that high PI turnover initiates apoptosis, however, has yet to be obtained.

It should be mentioned that overstimulation of T cells leading to cell death might occur late in the immune response, i.e. subsequent to the initial proliferative response. Thus, recent work has shown that, when Mls^a antigens are expressed on resting B cells, even high doses of these cells fail to elicit PI turnover of cloned T-cell lines even though strong proliferative responses occur [108; O'Rourke and Webb, unpubl. data]. By contrast, presentation of Mls^a antigens on activated B cells (which show higher levels of MMTV transcripts) does induce PI turnover. Since T-cell responses to Mls^a antigens are known to result in B cell activation, one can envision a scenario where continuous interaction between the responding T cells and Mls^a-bearing B cells results in a progressive increase in the overall strength (or avidity) of T-B-cell interaction and eventually to overstimulation of T cells. This notion of delayed onset of overstimulation might explain why the $V\beta 6^+$ T cells disappearing in Mls^a-injected mice apparently undergo several rounds of division before being eliminated.

The idea that high-avidity interactions between T cells and APCs eventually predispose to T-cell death predicts that cell death would be less prominent if the avidity of T-APC interactions were lowered. Indirect support for this prediction has come from the finding that the disappearance of Vβ6$^+$ cells in Mls^a-injected mice is influenced by the H-2 haplotype of the host [Webb, unpubl. data]. In the case of short-term proliferative responses to Mls^a antigens in vitro using limiting doses of APCs, Vβ6$^+$ T cells from I-E$^-$ H-2^b mice give appreciably lower anti-Mls^a responses than Vβ6$^+$ cells from I-E$^+$ H-2^k mice. The simplest explanation for the reduced Mls^a responsiveness of H-2^b Vβ6$^+$ T cells is that the binding avidity of these T cells for APCs expressing Mls^a/H-2^b is lower than for H-2^k Vβ6$^+$ cells responding to Mls^a/ H-2^k. This reduced avidity of H-2^b Vβ6$^+$ cells might be a reflection of determinant density and/or repertoire selection. Whatever the explanation, it is notable that the in vivo anti-Mls^a response in H-2^b and H-2^k mice is appreciably different. As discussed earlier, H-2^k mice show a strong initial proliferative response of Vβ6$^+$ cells following exposure to Mls^a cells; this response peaks on day 2–4 (40% of Vβ6$^+$ cells in spleen) and is then followed by a rapid and pronounced disappearance of Vβ6$^+$ cells [104]. In H-2^b mice, by contrast, the proliferation of Vβ6$^+$ T cells is less extensive (20–25% Vβ6$^+$ cells in spleen) and occurs more slowly, peak responses being delayed until day 4–6 [Webb, unpubl. data]. The subsequent elimination of Vβ6$^+$ cells is less marked: the level of these cells declines considerably but does not fall below the level in uninjected mice. Interestingly, the residual Vβ6$^+$ cells in these mice fail to mount in vitro proliferative responses to Mls^a antigens, implying that the cells are tolerant. At early stages after injection, tolerance seems to be a reflection of anergy. Thus, at 2–3 weeks after injection, the residual Vβ6$^+$ cells are refractory to stimulation with anti-Vβ6 antibody in vitro – a well-accepted criterion of anergy. At later stages (>2 months), however, the Vβ6 cells regain normal reactivity to anti-Vβ6 antibody, but remain poorly responsive to Mls^a antigens.

Our provisional interpretation of these findings on anti-Mls^a responses in H-2^b mice is as follows. As discussed above, we speculate that the less extensive deletion of Vβ6$^+$ cells in H-2^b than H-2^k mice reflects that T cells recognize Mls^a antigens with lower avidity in H-2^b mice: overstimulation is less prominent and fewer cells die. But what about the cells which evade death? Our suggestion is that these cells are a subset of low-avidity cells which are incapable of mounting overt proliferative responses to Mls^a antigens. When exposed to Mls^a antigens in vivo these cells undergo partial activation, which makes the cells refractory to TCR stimulation when subsequently tested in vitro. When the dose of antigen encountered in vivo falls below a certain threshold (most of the injected Mls^a lymphoid cells disappear within 2 weeks), the anergic cells revert to resting cells and thereby regain responsiveness to TCR stimulation. Because of their intrinsic low avidity for antigen, however, the recovered cells remain unresponsive to Mls[a] antigens when tested in vitro. It should be pointed out that it is unlikely that all of the cells displaying anergy eventually recover. Thus, when tested soon after initial stimulation in vivo, many of the cells typed as anergic in vitro are probably high-avidity cells in the process of being eliminated. Whether some of the low-avidity anergic cells are also eliminated is unclear.

The above discussion is highly speculative and it should be emphasized that the events which govern the fate of in vivo stimulated T cells are far from clear. Whether anergic cells recover or die is still unknown, and the notion that the strength of signal generated in T-APC interactions influences T-cell survival is more a theory than a fact. Nevertheless, the idea outlined above would seem to be a useful framework for designing more definitive experiments in the future.

As a final note, brief comment should be made on the relevance of the above results to the generation of T-cell memory. In this respect, the notion that T-cell contact with antigen in vivo leads to wide-scale elimination of the responding cells would seem to be incompatible with the dogma that secondary immune responses are more intense than primary responses. The point to emphasize here is that priming for strong secondary responses generally depends on initial priming with adjuvant. Since antigens suspended in adjuvant are retained in the immune system for prolonged periods [109], successful priming might hinge on T cells having continuous exposure to antigen. In support of this idea, recent work has shown that adoptively transferring 'memory' T cells in the absence of antigen causes the primed cells to rapidly disappear [110]. Memory, thus, appears to be short-lived in the absence of antigen. There is then no discrepancy with the finding that injecting Mls[a] lymphoid cells (without adjuvant) into Mls[b] hosts causes considerable elimination of Mls[a]-reactive host Vβ6[+] cells. In this situation antigen is cleared rapidly. The key question is whether prolonging T-cell contact with antigen in this model would prevent T cells from disappearing. This question is currently under investigation.

Concluding Comments

Although tolerance induction is largely a reflection of clonal deletion occurring in the thymus, it is clear that T-cell responses can also be regulated at the level of mature T cells in the periphery. In some situations, T cells may be down-regulated by other immune cells or products (e.g. inhibitory cytokines) released by these cells. In other situations, tolerance may be a reflection of inappropriate presentation of antigen, causing T cells to enter an anergic state. Tolerance can also be the end result of a powerful immune response, many of the cells participating in the response eventually being eliminated. This form of tolerance is especially prominent in the response to superantigens such as Mls antigens and bacterial toxins. We speculate that this process of deletion is largely restricted to high-avidity T cells. Extrathymic deletion of T cells could be a normal sequel to all immune responses and represent a mechanism for preventing overrepresentation of cells with a given specificity. Although contact with high doses of antigen could pose the risk of excessive deletion leading to functional tolerance, the continuous generation of new T cells in the thymus would limit this potential danger.

References

1 Festenstein H: Immunogenetic and biological aspects of in vitro lymphocyte allo-transformation (MLR) in the mouse. Transplant Rev 1973;15:62–88.
2 Janeway CA Jr, Lerner EA, Jason JM, Jones B: Lymphocytes responding to Mls-locus antigens are Ly-1⁺2⁻ and I-A restricted. Immunogenetics 1980;10:481–497.
3 Lutz CT, Glasebrook AL, Fitch FW: Enumeration of alloreactive helper T lymphocytes which cooperate with cytolytic T lymphocytes. Eur J Immunol 1981;11:726–734.
4 Miller RA, Stutman O: Enumeration of IL-2 secreting helper T cells by limiting dilution analysis and demonstration of unexpectedly high levels of IL-2 production per responding cell. J Immunol 1982;128:2258–2264.
5 Korngold R, Sprent J: Lethal graft versus host disease after bone marrow transplantation across minor histocompatibility barriers in mice. J Exp Med 1978;148:1687–1698.
6 Sachs JA, Huber B, Pena-Martinez J, Festenstein H: Genetic studies and effect on skin allograft survival of DBA/2, DAG, Ly, and M locus antigens. Transplant Proc 1973;5:1385–1387.
7 Lilliehöök B, Jacobsson H, Blomgrem H: Specifically decreased MLC response of lymphocytes from CBA mice injected with cells from the H-2-compatible, M-antigen-incompatible strain C3H. Scand J Immunol 1975;4:209–216.
8 Janeway CA Jr: Selective elements for the Vβ region of the T cell receptor: Mls and the bacterial toxic mitogens. Adv Immunol 1991;50:1–53.
9 Webb SR, Molnar-Kimber K, Bruce J, Sprent J, Wilson DB: T cell clones with dual specificity for Mls and various major histocompatibility complex determinants. J Exp Med 1981;154:1970–1974.
10 Festenstein H: Pertinent features of M locus determinants including revised nomenclature and strain distribution. Transplantation 1974;18:555–557.
11 Abe R, Hodes RJ: T cell recognition of minor lymphocyte stimulating (MLS) gene products. Annu Rev Immunol 1989;7:683–708.

12 Pullen AM, Marrack P, Kappler J: Evidence that Mls-2 antigens which delete Vβ3⁺ T cells are controlled by multiple genes. J Immunol 1989;142:3033–3037.

13 Abe R, Foo-Phillips M, Hodes RJ: Analysis of Mlsᶜ genetics: A novel instance of genetic redundancy. J Exp Med 1989;170:1059–1073.

14 Abromson-Leeman SR, Laning JC, Dorf ME: T cell recognition of Mlsᶜˑˣ determinants. J Immunol 1988;140:1726–1731.

15 Janeway CA Jr, Fischer-Lindahl K, Hammerling U: The Mls locus: New clues to a lingering mystery. Immunol Today 1988;9:125–127.

16 MacDonald HR, Schneider R, Less RK, Howe RC, Acha-Orbea H, Festenstein H, Zinkernagel RM, Hengartner H: T-cell receptor Vβ use predicts reactivity and tolerance to Mlsᵃ-encoded antigens. Nature 1988;332:40–45.

17 Kappler JW, Staerz U, White J, Marrack PC: Self tolerance eliminates T cells specific for Mls-modified products of the major histocompatibility complex. Nature 1988;332:35–40.

18 Kanagawa O, Palmer E, Bill J: The T cell receptor Vβ6 domain imparts reactivity to the Mls-1ᵃ antigen. Cell Immunol 1989;119:412–426.

19 Okada CY, Holzmann B, Guidos C, Palmer E, Weissman IL: Characterization of a rat monoclonal antibody specific for a determinant encoded by the Vβ7 gene segment: Depletion of Vβ7⁺ T cells in mice with Mls-1ᵃ haplotype. J Immunol 1990; 144:3473–3477.

20 Happ MP, Woodland DL, Palmer E: A third T cell receptor β chain variable region gene encodes reactivity to Mls-1ᵃ gene products. Proc Natl Acad Sci USA 1989;86: 6293–6296.

21 Webb SR, Sprent J: Response of mature unprimed CD8⁺ T cells to Mlsᵃ determinants. J Exp Med 1990;171:953–958.

22 Abe R, Vacchio MS, Fox B, Hodes RJ: Preferential expansion of the T cell receptor Vβ3 gene by Mlsᶜ reactive T cells. Nature 1988;335:827–830.

23 Fry AM, Matis LA: Self tolerance alters T cell receptor expression in an antigen-specific MHC restricted immune response. Nature 1988;335:830–832.

24 Pullen AM, Marrack P, Kappler JW: The T cell repertoire is heavily influenced by tolerance to polymorphic self antigens. Nature 1988;335:796–801.

25 Kappler JW, Roehm N, Marrack P: T cell tolerance by clonal elimination in the thymus. Cell 1987;49:273–280.

26 Tomanari K, Lovering E: T-cell receptor-specific monoclonal antibodies against a Vβ11-positive mouse T-cell clone. Immunogenetics 1988;28:445–451.

27 Bill J, Kanagawa O, Woodland DL, Palmer E: The MHC molecule I-E is necessary but not sufficient for the clonal deletion of Vβ11-bearing T cells. J Exp Med 1989; 169:1405–1419.

28 Woodland D, Happ MP, Bill J, Palmer E: Requirement for cotolerogenic gene products in the clonal deletion of I-E reactive T cells. Science 1990;247:964–967.

29 Pullen AM, Wade T, Marrack P, Kappler J: Identification of the region of T cell receptor β chain that interacts with the self superantigen Mls-1ᵃ. Cell 1990;61:1365–1374.

30 Cazenave P-A, Marche PN, Jouvin-Marche E, Voegtle D, Bonhomme F, Bandeira A, Coutinho A: Vβ17 gene polymorphism in wild-derived mouse strains: Two amino acid substitutions in the Vβ17 region greatly alter T cell receptor specificity. Cell 1990;63:717–728.

31 Frankel WN, Rudy C, Coffin J, Huber BT: Linkage of Mls genes to endogenous mammary tumor viruses of mice. Nature 1991;349:526–528.

32 Dyson PJ, Knight AM, Fairchild S, Simpson E, Tomonari K: Genes encoding ligands for deletion of Vβ11 T cells cosegregate with mammary tumor virus genomes. Nature 1991;349:531–532.

33 Woodland DL, Happ MP, Gollub KJ, Palmer E: An endogenous retrovirus mediating deletion of αβ T cells. Nature 1991;349:529–530.

34 Acha-Orbea H, Shakhov AN, Scarpellino L, Kolb E, Müller V, Vessaz-Shaw A, Fuchs R, Blöchlinger K, Rollini P, Billotte J, Sarafidou M, MacDonald HR, Diggelmann H: Clonal deletion of Vβ14-bearing T cells in mice transgenic for mammary tumor virus. Nature 1991;350:207–211.

35 Marrack P, Kushnir E, Kappler J: A maternally inherited superantigen encoded by a mammary tumor virus. Nature 1991;349:524–526.

36 Hugin AW, Vacchio MS, Morse HC III: A virus-encoded 'superantigen' in a retrovirus induced immunodeficiency syndrome of mice. Science 1991;252:424–427.

37 Imberti L, Sottini A, Bettinardi A, Puoti M, Primi D: Selective depletion in HIV infection of T cells that bear specific T cell receptor Vβ sequences. Science 1991;254:860–862.

38 Choi Y, Kappler JW, Marrack P: A superantigen encoded in the open reading frame of the 3′ long terminal repeat of mouse mammary tumor virus. Nature 1991;350:203–207.

39 Woodland DL, Lund FE, Happ MP, Blackman MA, Palmer E, Corley RB: Endogenous superantigen expression is controlled by mouse mammary tumor proviral loci. J Exp Med 1991;174:1255–1258.

40 Acha-Orbea H, Palmer E: Mls – A retrovirus exploits the immune system. Immunol Today 1991;12:356–361.

41 van Klaveren P, Bentvelzen P: Transactivating potential of the 3′ open reading frame of murine mammary tumor virus. J Virol 1988;62:4410–4413.

42 Salmons B, Erfle V, Brem G, Günzberg WH: *naf* a *trans*-regulating negative-acting factor encoded within the mouse mammary tumor virus open reading frame region. J Virol 1990;64:6355–6359.

43 Fleischer B, Schrezenmeier H: T cell stimulation by staphylococcal enterotoxins: Clonally variable response and requirement for major histocompatibility complex class II molecules on accessory or target cells. J Exp Med 1988;167:1697–1707.

44 Janeway CA Jr, Yagi J, Conrad PJ, Katz ME, Jones B, Vroegop S, Buxser S: T cell responses to Mls and to bacterial proteins that mimic its behavior. Immunol Rev 1989;107:61–88.

45 Marrack P, Kappler J: The staphylococcal enterotoxins and their relatives. Science 1990;248:705–710.

46 Cole BC, Atkin CL: The *Mycoplasma arthritidis* T-cell mitogen MAM: A model superantigen. Immunol Today 1991;12:271–276.

47 Tomai M, Kotb M, Majumdar G, Beachey EH: Superantigenicity of streptococcal M protein. J Exp Med 1990;172:359–362.

48 Matthes M, Schrezenmeier H, Homfeld J, Fleischer S, Malissen B, Kirchner H, Fleischer B: Clonal analysis of human T cell activation by the *Mycoplasma arthritidis* mitogen (MAS). Eur J Immunol 1988;18:1733–1737.

49 White J, Herman A, Pullen AM, Kappler JW, Marrack P: The Vβ-specific superantigen staphylococcal enterotoxin B: Stimulation of mature T cells and clonal deletion in neonatal mice. Cell 1989;56:27–35.

50 Rellahan BL, Jones LA, Kruisbeek AM, Fry AM, Matis LA: In vivo induction of anergy in peripheral Vβ8$^+$ T cells by staphylococcal enterotoxin B. J Exp Med 1990; 172:1091–1100.

51 Kawabe Y, Ochi A: Selective anergy of Vβ8$^+$ T cells in staphylococcus enterotoxin B-primed mice. J Exp Med 1990;172:1065–1070.

52 Kawabe Y, Ochi A: Programmed cell death and extrathymic reduction in Vβ8$^+$ CD4$^+$ T cells in mice tolerant to *Staphylococcus aureus* enterotoxin B. Nature 1991;349: 245–248.

53 Ikejima T, Dinarello CA, Gill DM, Wolff SM: Induction of human interleukin 1 by a product of *Staphylococcus aureus* associated with a toxic shock syndrome. J Clin Invest 1984;73:1312–1320.

54 Parsonnet J, Hickman RK, Eardley DD, Pier GB: Induction of human interleukin-1 by toxic-shock-syndrome toxin-1. J Infect Dis 1985;152:514–522.

55 Parsonnet J, Gillis ZA: Production of tumor necrosis factor by human monocytes in response to toxic-shock-syndrome toxin-1. J Infect Dis 1988;158:1026–1033.

56 Jupin C, Anderson S, Damais C, Alouf JE, Parant M: Toxic shock syndrome toxin 1 as an inducer of human necrosis factors and γ interferon. J Exp Med 1988;167:752–761.

57 Mourad W, Scholl P, Diaz A, Geha R, Chatila T: The staphylococcal toxic shock syndrome toxin 1 triggers B cell proliferation and differentiation via major histocompatibility complex-unrestricted cognate T/B cell interaction. J Exp Med 1989; 170:2011–2022.

58 Sharma S, King LB, Corley RB: Molecular events during B lymphocyte differentiation: Induction of endogenous mouse mammary tumor proviral envelope transcripts after B cell stimulation. J Immunol 1988;141:2510–2518.

59 King LB, Lund FE, White DA, Sharma S, Corley RB: Molecular events in B lymphocyte differentiation: Inducible expression of the endogenous mouse mammary tumor proviral gene, *mtv-9*. J Immunol 1990;144:3218–3227.

60 Steinman RM: The dendritic cell system and its role in immunogenicity. Annu Rev Immunol 1991;9:271–296.

61 Sprent J, Webb SR: Function and specificity of T cell subsets in the mouse. Adv Immunol 1987;41:39–133.

62 von Boehmer H, Sprent J: Expression of M locus differences by B cells but not T cells. Nature New Biol. 1974;249:363–365.

63 Ahmed A, Scher I, Smith AH, Sell KW: Studies on non H-2 linked lymphocyte activation determinants. J Immunogenet 1977;4:201–213.

64 Webb SR, Okamoto A, Ron Y, Sprent J: Restricted tissue distribution of Mls^a determinants. J Exp Med 1989;169:1–12.

65. Webb SR, Hu Li J, Wilson DB, Sprent J: Capacity of small B cell-enriched populations to stimulate mixed lymphocyte reactions: Marked differences between irradiated vs. mitomycin C-treated stimulators. Eur J Immunol 1985;15:92–96.

66 June CH, Ledbetter JA, Linsley PS, Thompson CB: Role of the CD28 receptor in T-cell activation. Immunol Today 1990;11:211–216.

67 Sunshine GH, Mitchell TJ, Czitrom AA, Edwards S, Glasebrook AL, Kelso A, MacDonald HR: Stimulator requirements for primed alloreactive T cells: Macrophage and dendritic cells activate T cells across all genetic disparities. Cell Immunol 1985;91:60–74.

68 Inaba M, Inaba K, Hosono M, Kamamoto T, Ishida T, Maramatsu S, Masuda T, Ikehara S: Distinct mechanisms of neonatal tolerance induced by dendritic cells and thymic B cells. J Exp Med 1991;173:549–559.

69 Metlay JP, Pure E, Steinman RM: Distinct features of dendritic cells and anti-immunoglobulin activated B cells as stimulators of the primary mixed leukocyte reaction. J Exp Med 1989;169:239–254.

70 Ramsdell F, Lantz T, Fowlkes BJ: A nondeletional mechanism of thymic self tolerance. Science 1989;246:1038–1041.

71 Spieser DE, Schneider R, Hengartner H, MacDonald HR, Zinkernagel RM: Clonal deletion of self reactive T cells in irradiation bone marrow chimeras and neonatally tolerant mice: Evidence for transfer of Mls[a]. J Exp Med 1989;170:595–600.

72 Fraser JD: High affinity binding of staphylococcal enterotoxins A and B to HLA-DR. Nature 1989;339:221–223.

73 Dellabona P, Peccoud J, Kappler J, Marrack P, Benoist C, Mathis D: Superantigens interact with MHC class II molecules outside of the antigen groove. Cell 1990;62:1115–1121.

74 Bierer BE, Sleckman BP, Ratnofsky SE, Burakoff SJ: The biologic roles of CD2, CD4 and CD8 in T-cell activation. Annu Rev Immunol 1989;7:579–599.

75 MacDonald HR, Lees RK, Chvatchko Y: CD8[+] T cells respond clonally to Mls-1[a] encoded determinants. J Exp Med 1990;171:1381–1386.

76 Gao E-K, Kanagawa O, Sprent J: Capacity of unprimed CD4[+] and CD8[+] T cells expressing Vβ11 receptors to respond to I-E alloantigens in vivo. J Exp Med 1989;170:1947–1957.

77 Fowlkes BJ, Schwartz RH, Pardoll DM: Deletion of self reactive thymocytes occurs at a CD4[+]8[+] precursor stage. Nature 1988;334:620–623.

78 MacDonald HR, Hengartner H, Pedrazzini T: Intrathymic deletion of self-reactive cells prevented by neonatal anti-CD4 antibody treatment. Nature 1988;335:174–176.

79 Blackman M, Kappler J, Marrack P: The role of the T cell receptor in positive and negative selection of developing T cells. Science 1990;248:1335–1341.

80 von Boehmer H, Teh HS, Kisielow P: The thymus selects the useful, neglects the useless and destroys the harmful. Immunol Today 1989;10:57–61.

81 Pircher H, Bürki K, Lang R, Hengartner H, Zinkernagel RM: Tolerance induction in double specific T-cell receptor transgenic mice varies with antigen. Nature 1989;342:559–561.

82 Sprent J, Gao E-K, Webb SR: T cell reactivity to MHC molecules: Immunity versus tolerance. Science 1990;248:1357–1363.

83 Webb SR, Sprent J: Tolerogenicity of thymic epithelium. Eur J Immunol 1990;20:2525–2528 (correction published 1991;21:534).

84 Webb SR, Sprent J: Induction of neonatal tolerance to Mls[a] antigens by CD8[+] T cells. Science 1990;248:1643–1646.

85 Mazda O, Watanabe Y, Gyotoku J-I, Katsura Y: Requirement of dendritic cells and B cells in the clonal deletion of Mls-reactive T cells in the thymus. J Exp Med 1991;173:539–547.

86 Agus DB, Surh CD, Sprent J: Reentry of T cells to the adult thymus is restricted to activated cells. J Exp Med 1991;173:1039–1046.

87 Schneider R, Lees RK, Pedrazzini T, Zinkernagel RM, Hengartner H, MacDonald HR: Postnatal disappearance of self-reactive (Vβ6[+]) cells from the thymus of Mls[a]

mice: Implications for T cell development and autoimmunity. J Exp Med 1989;169: 2149–2158.

88 Feeney AJ: Junctional sequences of fetal T cell receptor β chains have few N regions. J Exp Med 1991;174:115–124.

89 Gershon RK, Cohen P, Hencin R, Liebhaber SA: Suppressor T cells. J Immunol 1972;108:586–590.

90 Green D, Flood P, Gershon R: Immunoregulatory T-cell pathways. Annu Rev Immunol 1983;1:439–464.

91 Dorf M, Benacerraf B: Suppressor cells and immunoregulation. Annu Rev Immunol 1984;2:127–158.

92 Hodes RJ: T cell mediated regulation: Help and suppression; in Paul WE (ed): Fundamental Immunology. New York, Raven Press, 1989, pp 587–620.

93 Gajewski TF, Fitch FW: Anti-proliferative effect of IFN-gamma in immune regulation. I. IFN-gamma inhibits the proliferation of TH2 but not TH1 murine helper T lymphocyte clones. J Immunol 1988;140:4245–4252.

94 Florentino DF, Zlotnik A, Vieira P, Mosmann TR, Howard M, Moore KW, O'Garra A: IL-10 acts on the antigen presenting cell to inhibit cytokine production by TH1 cells. J Immunol 1991;146:3444–3451.

95 Jenkins MK, Pardoll DM, Mizuguchi J, Quill H, Schwartz RH: T cell unresponsiveness in vivo and in vitro: Fine specificity of induction and molecular characterization of the unresponsive state. Immunol Rev 1987;95:113–135.

96 Mueller DL, Jenkins MK, Schwartz RH: Clonal expansion versus functional clonal inactivation: A costimulatory signalling pathway determines the outcome of T cell antigen receptor occupancy. Annu Rev Immunol 1989;7:445–480.

97 Jenkins MK, Taylor PS, Norton SD, Urdahl KB: CD28 delivers a costimulatory signal involved in antigen specific IL-2 production by human T cells. J Immunol 1991;147:2461–2466.

98 Burkly LC, Lo D, Kanagawa O, Brinster RL, Flavell RA: T cell tolerance by clonal anergy in transgenic mice with nonlymphoid expression of MHC class II I-E. Nature 1989,342:564–566.

99 Miller JFAP, Morahan G, Allison J, Hoffman M: A transgenic approach to the study of peripheral T cell tolerance. Immunol Rev 1991;122:103–116.

100 Hammerling GJ, Schourich G, Momburg F, Auphan N, Malissen M, Malisson B, Schmitt-Verhulst A-M, Arnold B: Nondeletional mechanisms of peripheral and central tolerance: Studies with transgenic mice with tissue-specific expression of a foreign MHC class I antigen. Immunol Rev 1991;122:47–67.

101 Zinkernagel RM, Pircher HP, Ohashi P, Oehen S, Odermatt B, Mak T, Arnheiter H, Bürki K, Hengartner H: T and B cell tolerance and responses to viral antigens in transgenic mice: Implications for the pathogenesis of autoimmune versus immunopathological disease. Immunol Rev 1991;122:133–171.

102 Blackman MR, Gerhardt-Burgert H, Woodland DL, Palmer E, Kappler J, Marrack P: A role for clonal inactivation in T cell tolerance to Mls-1^a. Nature 1990;345:540–542.

103 Rammensee H-G, Kroschewski R, Frangoulis B: Clonal anergy induced in mature Vβ6$^+$ T lymphocytes on immunizing Mls-1^b mice with Mls-1^a expressing cells. Nature 1989;339:541–544.

104 Webb SR, Morris C, Sprent J: Extrathymic tolerance of mature T cells: Clonal elimination as a consequence of immunity. Cell 1990;63:1249–1256.

105 Rocha B, von Boehmer H: Peripheral selection of the T cell repertoire. Science 1991;
     251:1225–1228.
106 Jones LA, Chin LT, Longo DL, Kruisbeek AM: Peripheral clonal elimination of
     functional T cells. Science 1990;250:1726–1729.
107 Sprent J: Fate of H-2 activated T lymphocytes in syngeneic hosts. I. Fate in lymphoid
     tissues and intestines traced with ^{3}H-thymidine, ^{125}I-deoxyuridine and [51]chromium.
     Cell Immunol 1976;21:278–302.
108 O'Rourke AM, Mescher MF, Webb SR: Activation of polyphosphoinositide hydroly-
     sis in T cells by H-2 alloantigen but not Mls determinants. Science 1990;249:171–
     174.
109 Szakal AK, Kosco MH, Tew JG: Microanatomy of lymphoid tissue during humoral
     immune responses. Annu Rev Immunol 1989;7:91–109.
110 Gray D, Matzinger P: T cell memory is short lived in the absence of antigen. J Exp
     Med 1991;174:969–974.

Susan R. Webb, PhD, Department of Immunology, IMM4A,
The Scripps Research Institute, 10666 North Torrey Pines Road,
La Jolla, CA 92037 (USA)

Fleischer B (ed): Biological Significance of Superantigens.
Chem Immunol. Basel, Karger, 1992, vol 55, pp 115–136

Effects of Staphylococcal Toxins on T-Cell Activity in vivo

Atsuo Ochi, Kouichi Yuh, Kiyoshi Migita, Yojiro Kawabe

Division of Neurobiology and Molecular Immunology, Samuel Lunenfeld
Research Institute, and Department of Immunology and Medical Genetics,
University of Toronto, Mount Sinai Hospital, Toronto, Ont., Canada

Introduction

Enterotoxins, a common cause of food poisoning, are proteins produced
by the staphylococci. These proteins are classified according to their reaction
with specific antibodies as enterotoxin A, B, C, etc. [1]. The molecular
structure of these enterotoxins is well established and the amino acid and
DNA sequences of each toxin revealed that they are members of a closely
related family. A powerful mitogenic capacity for T lymphocytes of several
species by these proteins was reported [2]. T-cell activation by staphylococcal
enterotoxins (SEs) was subsequently found to be restricted to accessory cells
which express class II major histocompatibility complex (MHC) [3, 4].
Biochemical characterization of class II MHC dependency of T-cell stimulation demonstrated binding affinities for these proteins [5–7]. The toxins were
not acting as mitogens such as concanavalin A (Con A). That is, the toxins do
not activate in vitro as large a proportion of T cells as the stimulatory plant
lectins and some T-cell clones could not be stimulated by particular toxins
[3]. The mechanism of selective reactivity of toxins to particular T-cell clones
was discovered by White, who demonstrated the close association of T-cell
receptor (TCR) Vβ usage and reactivity to staphylococcal enterotoxin B
(SEB) [8]. A number of studies reported similar results and the Vβ specificity
of these toxins was demonstrated in both mice and humans [9]. These
characteristics have striking similarities to the minor lymphocyte stimulating
(Mls) antigens which were discovered by Festenstein in 1973 [4, 10].
However, the Mls antigen is an endogenously inherited cellular component

116

unlike the exocrine substances of bacterias [4, 11]. These bacterial enterotoxins and Mls antigens were labeled 'superantigens' because of the potential to trigger strong activation in a large proportion of peripheral T cells. A major role of Mls antigen in thymic tolerance by deletion (negative selection) of immature T cells carrying particular Vβs TCR was documented [12–14]. The paradigm of Mls antigen which causes profound effects on the development of the T-cell repertoire in the thymus and on T-cell function in the periphery encouraged us to take a similar approach using bacterial superantigens. However, the experimental system using bacterial superantigens is likely far more quantitative than that using cellularly expressed Mls antigen where the structure and amount of the 'antigen' are unknown.

In this article we will discuss the in vivo effects of a bacterial superantigen on T cells in mice. The sections will present topics concerned with the activation induced proliferation, activation induced programmed cell death (PCD) and functional unresponsiveness (anergy) of mature T cells in the periphery.

In vivo Activation and Death of T Cells with SEs

SEs strong activation of T cells expressing limited Vβs TCRs has been studied in vitro in mice and humans. The activation of such T cells induces the production of lymphokines such as interleukin-2 (IL-2), gamma-interferon (γIFN), tumor necrosis factor (TNF) and lymphotoxins [15, 16]. Both CD4 and CD8 subsets proliferate to SEB [17, 18]. In contrast to CD4 T cells, which require a signal from CD4 molecules, CD8 T-cell activation by SEB is independent of CD8 molecules [18]. Activated T cells exert cytotoxicity to SEB-presenting class II MHC+ B lymphoma target cells [17]. Cytotoxicity was observed in CD4 T cells but not in CD8 T cells when fresh mouse spleen T cells were stimulated [17]. Cytotoxicity similar to that of a lymphokine-activated killer (LAK) cell was also observed in SEA-activated human peripheral blood mononuclear cells [19]. Therefore, SEs activation of mature T cells is similar to anti-CD3 activated T cells and triggers almost a full range of functions in reactive T cells.

In vivo effects of bacterial superantigens were previously studied in regard to toxic shock syndrome (TSS) which takes place in association with gram-positive bacterial infections. Experimental enterotoxemia with the main symptoms of emesis and diarrhea could be produced in monkeys or cats by intravenous or oral administration of these bacterial superantigens

[2, 20–22]. Lymphocytosis in peripheral blood and a reduction of serum immunoglobulins were observed in TSS. The immunosuppressive effect of SEB, which reduced the number of specific plaque-forming B cells in sheep red blood cell-primed mice, was also reported [23]. Whereas the initial T lymphocytosis in the periphery in response to SEs was appreciated for a long time, it is only very recently that the fate of these activated cells in terms of the development of peripheral tolerance was appreciated.

In vivo effects of bacterial superantigens to T cells have been best studied using SEB, which causes strong activation in T cells expressing Vβ7, 8.1,2,3 TCRs in mice [9]. In neonatal mice injected with SEB every other day, immature and mature Vβ8+ T cells were deleted in both CD4 and CD8 subsets in thymus [8]. When adult mice were SEB-primed by intravenous (i.v.), intraperitoneal (i.p.) or subcutaneous routes, the yield of thymocytes decreased [24]. However, the proportion of Vβ8+ T cells in the periphery was elevated to approximately twice that of normal mice by day 3 after injection [25–27]. This expansion of Vβ8+ T cells in the periphery occurred in both CD4 and CD8 subsets. By the end of the first week, the number of Vβ8+ T cells declined to a level below normal in CD4+ T cells [25, 26]. The decline of Vβ8+ T cells in the CD8 subset was less than that in CD4 subset and returned to that of the normal animal [25, 26]. In conjunction with the reduction of Vβ8+ T cells in vivo, a study of the DNA sample of spleen T cells revealed they were undergoing PCD [25]. The expansion, reduction and PCD took place specifically in Vβ8+ T cells but not in Vβ6+ or Vβ2+ T cells which do not react to SEB [25, 26]. The induction of IL-2 receptor (IL-2R) expression on Vβ8+, CD4 and CD8 T cells occurred within 18 h after SEB injection [27]. Expression of CD8 was enhanced at the same time in CD8+, Vβ8+ T cells [27]. These SEB-'primed' cells, when tested within 24 h after priming, proliferated in response to IL-2 or TCR stimulation (SEB and allogenic stimulations) more efficiently and with quicker kinetics than normal cells [26, 27]. The level of activation as measured by IL-2R expression and proliferation, depended on the dose of SEB injected. However, the expansion of Vβ8+ T-cell proportion was usually less than 300% and the reduction hardly reached below 50% of normal level, even at high doses (~ 800 μg) [unpubl. results]. γIFN was induced in vivo and detected in the serum of SEB-primed mice 10 h after injection [27]. These cells, activated in vivo, were suggested to contribute to the prevention of outgrowth of a malignant tumor which was inoculated with SEB concomitantly [27]. By day 4 after SEB injection, spleen cells lost most of the responsiveness (~ 80%) to SEB despite the fact that spleen had similar or more numbers of Vβ8+ T cells than normal mice [26, 27].

SEs in other animal systems have given slightly different responses. Studies of peripheral blood lymphocytes from monkeys injected with a small amount of SEA (0.5 μg/kg of body weight), demonstrated transient leukopenia lasting approximately 24 h [23]. The leukopenia converted to lymphocytosis by 48 h after injection. However, SEA-primed monkeys were not unresponsive to SEA 16 days after injection. This also contrasted to the efficient induction of SEA-specific unresponsiveness on day 12 in 10 μg SEA-injected mice [26]. This may indicate the requirement of a high dose of bacterial superantigen to induce unresponsiveness. Alternatively, the consequence of T-cell stimulation by a bacterial superantigen may be variable betwen species. An example was reported in rats where SEB behaved more like a suppressant than a stimulant [28]. In summary, the administration of bacterial superantigen causes proliferation, lymphokine production, activation induced death and unresponsiveness of specific T cells in vivo.

Programmed Cell Death Occurs in Vβ8+ Cells Activated by SEB

In order to understand the cellular events that underlie SEB administration and the resulting activation and deletion of T cells in mice in detail, Balb/c mice were injected i.p. with ^{125}I-labeled SEB. Figure 1 shows the result of blood radioactivity kinetics assayed in ^{125}I-labeled SEB-primed mice. The blood's peak of radioactivity appeared 1 h after the administration. The radioactivity declined quickly within another hour after this peak. Then the radioactivity of the blood dropped gradually and returned to the background level within 24 h of SEB injection. The analysis of SEB-injected mouse sera by SDS-polyacrylamide gel fractionation and autoradiography also showed varying intensities of SEB bands which matched the radioactivity of blood samples (data not shown). The data suggested that i.p.-injected SEB was quickly incorporated into and cleared from the blood stream in mice. Whereas injected SEB was cleared from the blood within 24 h, about 3% of the injected radioactivity remained in the spleen 24 h later. However, spleen cells prepared from mice at this time did not stimulate co-cultured fresh normal spleen T cells (data not shown). These results indicate that injected SEB remained in circulation for a relatively short time and even that deposited in spleen was not able to induce T-cell stimulation after 24 h.

We next monitored in detail the activation-induced expansion and reduction of Vβ8+ T cells after administration of SEB (fig. 2). We mainly examined CD4+, Vβ8+ T cells which had a greater reduction in number after

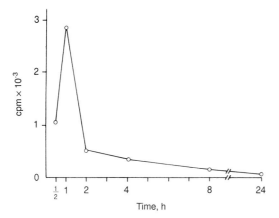

Fig. 1. Kinetics of radioactivity in blood in ^{125}I-SEB primed mice. Mice were injected with ^{125}I radiolabeled SEB (2×10^5 cpm/mouse). Blood was sampled at the indicated time points. The radioactivity of the blood (50 µl) was measured by γ-counter. Data indicate cpm value per ml of blood. Data represent one of the 5 individual mice samples which showed equivalent results.

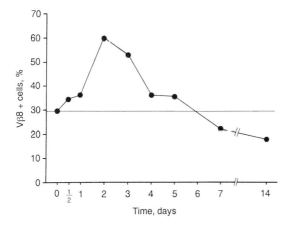

Fig. 2. Changes in proportion of Vβ8+ cells in the CD4+ T-cell subset over time in SEB-primed mice. Splenic cells from SEB-primed mice were studied for Vβ8 and CD4 expression at the indicated time points after SEB priming. Each result shown represents the mean of values obtained on three individual mice. The horizontal line indicates the proportion of Vβ8+ cells in the CD4+ T-cell subset in control mice (30%).

stimulation than CD8+, Vβ8+ T cells. Elevation of CD4+, Vβ8+ T-cell proportion was detectable 12 h after injection. The increase of Vβ8+ T cells was weak for the first 24 h. The most remarkable proliferation occurred between 24 and 48 h after SEB injection. By 48 h, CD4+, Vβ8+ T cells had doubled, after which the proportion started declining. The largest drop occurred between days 3 and 4 when the proportion returned to a level slightly above that of normal mice. After this, the proportion of CD4+, Vβ8+ T cells fell gradually to 50–60% of control by 14 days after SEB injection. The spleen size changed in parallel to the expansion of Vβ8+ T cells in vivo as the size enlarged more than twice that of control spleens at 48 h and the number of the cells per spleen was also increased correspondingly. Single cell suspensions of enlarged spleens contained a large number of T-cell blasts. However, after a week, the size of the spleen returned to normal. The data indicated that administrated SEB caused a transient proliferation of CD4+, Vβ8+ T cells in vivo between 24 and 48 h. It was also shown that the reduction of CD4+, Vβ8+ T cells occurred in two phases. One was the quick drop between 48 and 96 h and another was the gradual decline after 96 h which continued to the end of the second week. The proportion of CD4+, Vβ8+ T cells was stable for 2 months after this time.

The results obtained in the present study showed that the reduction of Vβ8+ T cells took place only after their initial expansion and suggested that PCD occurs in proliferating cells and depends on cellular activation signals. Studies on PCD in T-cell hybridomas have also linked this phenomenon with cell activation [29, 30]. To test that deletion occurred in activated cells in our system, cell death was compared between density gradient separated blastic and nonblastic spleen cells obtained from SEB-primed mice 3 days after SEB injection. DNA fragmentation was observed exclusively in the blastic cell fraction (fig. 3) as is consistent with the contention that cells targeted for clonal deletion undergo activation and blastogenesis prior to their PCD in vivo.

A Signal by the CD8 Molecule Enhances SEB-Induced
CD8+, Vβ8+ T Cell Reduction

The bacterial superantigens do not require processing to stimulate T cells. They bind outside of class II MHC molecules unlike protein antigens, which are presented in the antigen groove of class II MHC as a processed peptide [31, 32]. Although the superantigens are restricted to class II MHC molecules, they stimulate both CD4 and CD8 T cells. Anti-CD4 antibodies can inhibit activation of CD4 T cells by SEB whereas anti-CD8 antibodies do

Fig. 3. DNA fragmentation in enriched splenic T-cell blasts from SEB-primed mice. Splenic T cells obtained 3 days after SEB priming were separated into blastic and nonblastic populations by discontinuous density gradient. Discontinuous density gradients were prepared using isotonic separation medium (NycoPrep, 1.077 Animal, Nycomed Pharma AS, Oslo, Norway). Two dilutions of this medium (92 and 80%) were made by mixing with PBS (0.091 M, pH 7.2). Spleen T cells were resuspended in 2 ml of 80% solution, then overlayed on the top of discontinuous gradient layers of 100 and 92%, 2 ml each in 15 ml disposable tube. Generated gradients were centrifuged for 15 min at 800 g at 4 °C. At equilibrium after centrifugation distinct bands of cells can be observed at the interphases, and dead cells and debris at the very top of the gradient. Cells harvested from 92/80% interphase contains large number of blast (more than 70% of cells). Small nonblast cells are harvested from 100/92% interphase. DNA fragmentation was assayed in 2×10^6 blast (b) and nonblast (a) cells.

not. Since the soluble form of the TCR β-chain directly binds the complex of class II MHC and bacterial superantigen in the absence of CD4 [33], it is likely that the CD4 molecule enhances the activation of CD4 T cells by an additional signal [34]. CD8 T cells are activated by SEB, in the absence of a co-receptor molecule, as efficiently as are CD4 T cells, as assessed using a proliferation assay. In vitro and in vivo expansion of Vβ8+ T cells are similar for both CD4 and CD8 subsets. However, the in vivo deletion of Vβ8+ T cells in the CD4 subset was more remarkable than in the CD8 subset. The implication of this difference may be that CD8 T cells were activated differently in the absence of co-receptor molecules compared with CD4 T cells. In other words, a signal mediated by co-receptors may play an important role in triggering PCD in Vβ8+ cells in vivo. In order to test this hypothesis, a monoclonal antibody specific for SEB was generated in our

Table 1. Enhanced reduction of SEB stimulated CD8+, Vβ8+ T cells by SEB-CD8 cross-link in vivo

T cell subsets	% proportion[a]			
	control[b]	SEB[c]	Ab[d]-conjugate alone[e]	Ab-conjugate+SEB[f]
CD4, Vβ8	30	18	32	21
CD8, Vβ8	33	29	37	11

Data represent one of the three separate experiments.
[a] Proportion of Vβ8+ cells among CD4+ or CD8+ T cells.
[b] 0.2 ml PBS was injected.
[c] 50 μg SEB was injected.
[d] Anti-CD8 antibody (Ab) and anti-SEB Ab.
[e] 50 μg of protein was injected.
[f] 50 μg of Ab conjugate and 50 μg of SEB were injected.

laboratory from SEB-primed Balb/c mice. Antibodies were screened for SEB binding by ELISA using SEB coated plates. An antibody that had reactivity to SEB but did not prevent the activation of Vβ8 T cells by SEB-pulsed APC was selected. This antibody and anti-CD8 antibody were chemically coupled to create a bispecific construct. Bispecific antibody was expected to cross-link the CD8 molecule close to the TCR-SEB-class II MHC complex. We examined if SEB-induced deletion of CD8+, Vβ8+ T cells could be enhanced in vivo by co-injecting this bi-functional reagent with SEB. As shown in table 1, the proportion of CD8+, Vβ8+ T cells dramatically decreased when SEB and this reagent were injected concomitantly but not when injected with antibody complex alone. The CD4+, Vβ8+ T-cell response was not affected in the same spleen sample. The data strongly suggested that the activation-induced deletion of mature T cells was under the control of a co-receptor molecule mediated signal (fig. 4).

Accelerated Induction of PCD of SEB-Primed Spleen T Cells by Protein Synthesis Inhibitor in vitro

Apoptosis of SEB-activated Vβ8+ T cells started on day 3 after SEB injection. This Vβ8+ T cell specific death was accelerated by a protein synthesis inhibitor, cycloheximide (CH), at 37 °C, after 20 h in culture [26]. This may indicate that these in vivo PCD triggered cells actively resist DNA degradation and cell destruction by a mechanism which depends on ongoing de novo synthesis of protein. In order to test this hypothesis, spleen T cells

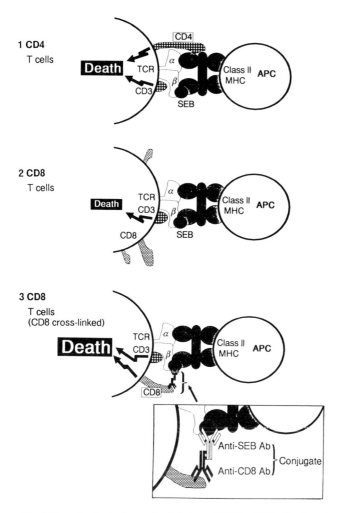

Fig. 4. Hypothetical scheme for deletion of CD8+, Vβ8+ T cells enhanced by a signal of CD8-TCR cross-linking (see text for discussion).

from days 1, 2 and 3 SEB-primed mice were cultured in vitro with 5 µg/ml CH and assayed for DNA fragmentation (fig. 5). The data showed fragmentation of DNA samples when prepared from 24- and 48-hour SEB-primed spleen T cells cultured in CH-containing medium. If, however, CH was not present in the culture, DNA fragmentation was barely detectable. When SEB-primed spleen T cells after 72 h were cultured with CH, the intensity of

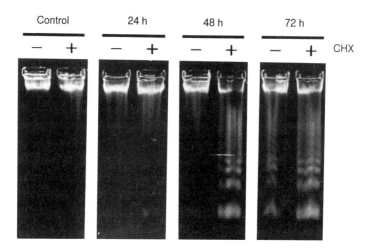

Fig. 5. CH induces DNA fragmentation in vitro in splenic T cell from SEB-primed mice. Splenic T cells were obtained from SEB-primed mice at 24, 48 and 72 h after SEB priming, cultured at 2×10^6/ml and at 37 °C for 90 min in the presence (+) or absence (−) of 5 μg/ml CH and the cultures then assayed for fragmentation. Representative examples of five separate experiments are shown.

fragmentation was enhanced. This effect by CH was dose-dependent and was observed when the concentration was above 0.01 μg/ml. The results indicated that SEB-stimulated cells resisted PCD by synthesizing a certain protein factor. It is possible that SEB-stimulated and proliferating cells are simply cell-cycle blocked by CH, which initiates DNA fragmentation. When mouse T- and B-cell hybridomas were incubated with CH for 90 min in vitro, there was no induction of DNA fragmentation, which is evidence against this possibility [30, and our personal observation]. A similar observation on DNA-degradation induction by CH in studies of B-chronic lymphocytic leukemic lymphocytes has been reported [35]. The conclusion is that activated Vβ8+ T cells are programmed to die relatively early after-SEB injection and that these cells were actively attempting to resist the death process.

The apparent failure of CH to abrogate SEB-induced death of Vβ8+ T cells in vitro should not necessarily be construed as implying that expression of PCD does not require de novo protein synthesis, as this assumption contradicts the general consensus which links de novo protein synthesis to PCD [36]. The interpretation of this discrepancy may be that these cells induced apoptosis by CH in culture because they carry ongoing PCD which had been triggered in vivo, and de novo protein synthesis may be required for the initiation of PCD. However, the process of DNA degradation and cell destruction may be independent from protein synthesis. The reason for this apparent discrepancy, of course, must wait for the results from detailed study of these Vβ8+ T cells triggered to death.

PCD represents an integral facet of many biological phenomena, but at present the mechanisms that trigger this type of cell death are largely unknown. With regard to PCD in T cells, recent data suggest that this process is triggered by cell-activation signals and possibly by virtue of negative feedback from secreted lymphokines [30, 37–39]. In one study, for example, stimulation of T-cell hybridomas via the TCR in vitro was found to induce not only cell activation, but also the initiation of signaling events which led to cell-cycle block, apparently by independent mechanisms [29, 40]. However, a causal link between T-cell activation and induction of PCD was not entirely obvious in the context of the in vivo model of tolerance examined here. Under normal circumstances, activation of T cells occurs very rapidly after engagement of the TCR, as indicated, for example, by the finding of a fiftyfold increase in expression of mRNA for the IL-2R, an activation marker, within 2 h of T-cell stimulation [41]. Similarly, membrane expression of IL-2R can be detected on 70% of Vβ8+ T cells within 18 h after their stimulation by SEB [27, and our personal observation]. In contrast, we were unable to detect PCD in the time period immediately following SEB injection, but instead observed cell death within the SEB-responsive Vβ8+ T-cell population only after these cells underwent a day of proliferation. These observations suggest that the commitment to PCD may be more directly related to cell proliferation.

A conflict has arisen concerning whether lymphokine starvation of activated cells plays a major role in PCD in causing SEB-stimulated Vβ8+ T cells [42, 43]. The PCD of IL-2-dependent T-cell clone in IL-2 free culture medium in vitro was reported previously [44]. Perhaps SEB-activated T cells become IL-2-dependent in vivo and discontinued production of lymphokines triggers the death of activated spleen T cells. This seems unlikely because 2 days after SEB priming, the percentage of Vβ8+ T cells in the periphery was elevated from 20 to 40% both before and after 20 h incubation at 37 °C, and anti-IL-2 antibodies added to this culture to block the possible effect of autocrine IL-2 did not induce PCD of Vβ8+ T cells [unpubl. observation]. Therefore, our results do not favor the contention of a key role for T-cell lymphokine starvation per se in the induction of cell death after SEB-priming.

Anergy Induction in Cells with SEs

The functional unresponsiveness of T lymphocytes has been reported by Schwartz's group using Th cell clones which were antigen-stimulated in vitro with (a) chemically fixed APC, and (b) antigen and purified class II MHC molecules inserted into planar membranes, or immobilized CD3 complex-specific antibodies [45–49]. These experiments implied that antigen-specific unresponsiveness of Th cells is the outcome of occupancy of TCR in the absence of co-stimulatory signals [45, 50]. The anergic Th cell clones expressed unchanged levels of TCR and IL-2R, and were responsive to exogenous IL-2. Antigen stimulation could not induce autocrine production of IL-2 because of defective transcription of the IL-2 gene in these tolerant cells.

The existence of anergic T cells in vivo was first demonstrated in mice tolerant to Mls antigens [51]. The stimulation of these cells with Mls antigen only induced marginal proliferation and IL-2 production in vitro. The study of allo-Mls antigen-induced T cell response in vivo was furthered by Webb et al. [52] who demonstrated clonal deletion as well as clonal anergy of reactive T cells in tolerant animals. Their data indicated that the activation-induced clonal deletion took place within 7 days after injection of Mls disparate cells.

Similar to Mls antigens, the administration of bacterial superantigens induced specific T cell anergy in vivo. Mice injected i.p. or i.v. with SEB did not specifically respond when spleen or LN cells were studied 7 days after SEB challenge [25, 26, 53]. Similar clonal anergy was also induced in SEA-injected mice [26]. Anergy was measured in both proliferation and IL-2 production assays. The degree of anergy depended on the dose of SEB and became evident when the SEB dose was above 1 µg/mouse. Results were equivalent even when mice were challenged with SEB, emulsified in complete Freund's adjuvant [53]. A detailed study of T cells in SEB-primed spleen demonstrated that only CD4 T cells and not CD8 were unresponsive [17]. Similar data which showed selective anergy of CD4 T cells was reported in a study of Mls-1[a] mice expressing a transgenic Vβ8.1 beta-chain [54]. Although these anergic Vβ8+ T cells were able to express IL-2R, they failed to proliferate in response to exogenous IL-2 [46, 52]. Therefore, in vivo induced clonal anergy of T cells by superantigens appears to be different from that induced in Th cell clones in vitro in response to IL-2.

To obtain more insight into the TCR-mediated signaling defect in SEB-induced anergic T cells, we measured the net response by CD4+, Vβ8+ T cells to various reagents (fig. 6). To this end, CD8 cell-depleted spleen cells were

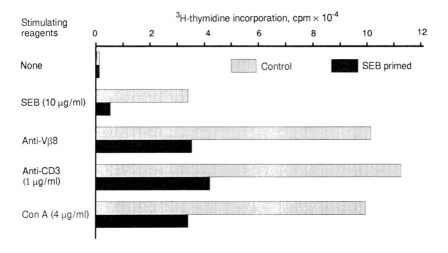

Fig. 6. Unresponsiveness of SEB-primed CD4+, Vβ8+ T cells. Spleen cells depleted of CD8+ cells were cultured in 24-well culture plates at 10^6 cells in 1 ml per well in the presence of SEB, Con A, anti-Vβ8 and anti-CD3 antibodies. After 40 h of culture cells were pulsed with [³H]-thymidine then Vβ8+ T cells were purified by magnet activated cell sorting technique to measure the incorporation of [³H]-thymidine. Results were average of quadruplicated samples. Control: Cells were prepared from PBS treated mice. SEB primed: Cells were prepared from 7-days 50-μg SEB-primed mice. Anti-Vβ8: Cells were stimulated with immobilized antibodies in culture plates.

cultured in vitro in the presence of SEB, anti-CD8, anti-CD3, or Con A for 48 h and [³H]-thymidine pulsed for the last 8 h. After culture, CD4+, Vβ8+ T cells were purified using a magnetically activated cell sorter (MACS) and a fixed number of cells were assayed for the incorporation of [³H]-thymidine. None of the reagents stimulated SEB-primed CD4+, Vβ8+ T cells to the same degree as normal CD4+, Vβ8+ T cells. However, the responses of cells stimulated with antibodies or Con A were better than those re-stimulated with SEB. Furthermore, generally, the responses of fresh anergic T cells appeared higher than those reported for anergic Th clones where proliferative responses were less than 10% of control response [44–49]. This difference may indicate the heterogeneity of Vβ8+ T cells existing in spleen, i.e. different affinities of cells to SEB or the possible existence of SEB-nonreactive Vβ8+ T cells which still respond to non-specific mitogens as it was suggested recently [55]. Alternatively, the unresponsiveness established in Th cell clones could be more strict than that of SEB-tolerant CD4+, Vβ8+

T cells. This difference may suggest the existence of multiple mechanisms that can result in 'shallow' and 'profound' T-cell unresponsiveness since, as mentioned, superantigen induced fresh anergic T cells and protein antigen-specific Th-cell clones also differ in IL-2 responsiveness. Because the mechanism of T cell activation by superantigens has a number of striking differences from classical protein antigens, which include trans-signaling pathways [56, 57], it might be the case that induction of anergy via superantigens uses a different pathway from that used by classical antigens.

Proliferation of anergic CD4+, Vβ8+ T cells was restored to normal levels by co-stimulation with phorbol-12-myristate-13-acetate (PMA) and ionomycin [personal observation]. Blackman et al. [58] reported altered antigen signaling in anergic T cells of self Mls antigen-specific T-cell receptor β-chain transgenic mice. Co-stimulation with PMA and ionomycin restored the proliferation in these cells. Therefore, these results seemed to indicate that superantigen-induced anergic T cells were hampered in their response primarily by an abnormality in a trans-signaling mechanism.

Protein Synthesis Inhibitor Prevents Anergy Induction in vivo

Whereas the molecular change underlying the anergy of T cells is not yet clear, a report showed that de novo protein synthesis was essential for T-cell unresponsiveness in vitro [47]. In order to understand the molecular mechanism of SEB-specific CD4+, Vβ8+ T cell anergy in vivo we have tested the effect of CH on the tolerance induction. A sublethal dose of CH was injected i.p. along with SEB in mice at different times. Tolerance to SEB stimulation was examined after 10 days (fig. 7). The results demonstrated that in vivo tolerance induction to SEB was dependent upon de novo protein synthesis in the first 12 h. After that period, injection of CH was unable to prevent anergy. CH treatment inhibited protein synthesis for 6 h in vivo as determined by measuring the time required for resumption of protein synthesis in spleen after pulsing mice with [35S]-methionine. Therefore, the protein synthesis important for induction of anergy seemed to take place within 12 h after SEB injection. It is not clear if T-cell anergy induction is prevented because of protein-synthesis inhibition in responding T cells or in other cellular components like APC. However, SEB does not require processing by APC to stimulate T cells [4] and TCR β-chain spontaneously binds to the high-affinity complexed SEB and class II MHC. Therefore, the inhibition of anergy by CH is most probably because of the inhibition of a protein synthesis essential for anergy in CD4+, Vβ8+ T cells in vivo.

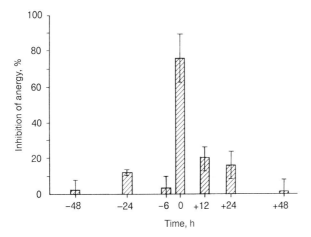

Fig. 7. Kinetics of CH-induced inhibition of SEB-specific anergy. Balb/c mice were injected with 1.0 mg CH i.p. at the indicated times (shown in hours) before and after SEB priming (zero-time point). Splenic cells suspensions prepared 7 days after SEB priming were cultured in 96-well plates at $2^3 \times 10^4$ cells per well and were stimulated in vitro with 10 μg/ml SEB. The percent inhibition of anergy was calculated as $100 \times$ [1 − (normal spleen response − SEB-primed and CH-treated spleen response)/(normal spleen response − SEB alone primed spleen response)]. The results represent the mean of values and standard deviations obtained from three individual mice in separate experiments.

Unresponsiveness of CD4+, Vβ8+ T Cells is Reversible in vivo

In order to investigate if anergic lymphocytes represent terminally differentiated cells, we have induced SEB-specific T-cell anergy in thymecto-mized mice and monitored the response to SEB at different times (fig. 8). As in normal mice, thymectomized mice became tolerant and the proportion of CD4+, Vβ8+ T cells was reduced to 60% of control in spleen when examined on day 10 after SEB priming. The response to SEA was comparable in both PBS-primed and 50-μg SEB-primed mice. When spleen cells were assayed in vitro for their SEB-specific proliferative response 2 months after SEB priming, the response was higher than 10 days after SEB priming but was still below that of control spleen cells. Thus, mice were apparently recovering from tolerance 2 months after SEB challenge. Four months later SEB-primed spleen cells responded efficiently to SEB and the cell dose-dependent res-ponse curve was equivalent to that of control spleen cells. IL-2 production of SEB-stimulated spleen cultures also recovered to control levels. The recovery of the proliferative response to SEB was further demonstrated when CD4+,

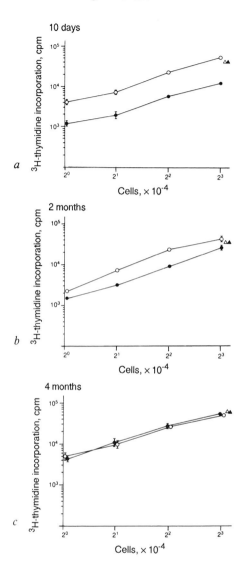

Fig. 8. SEB-specific spleen cell response of SEB-tolerated thymectomized mice on day 10, and after 2 and 4 months. Varied doses of spleen cells (horizontal axe) were cultured in the presence of 10 μg/ml SEB (circles) or 2.5 μg/ml SEA (triangles) 10 days, 2 months and 4 months after SEB-priming. Open circles and a triangle = Control spleen cells; black circles and a triangle = SEB-primed spleen cells.

Vβ8+ T cells of SEB-primed and control mice were purified and their response was determined. The proliferation and IL-2 production to varied doses of SEB were identical between previously stimulated and normal cell samples. The percentage of Vβ8+, CD4+ or CD8+ T cells on day 10 was maintained even after 4 months and there was no increase in the number of CD4+, Vβ8+ spleen T cells (data not shown).

Our results showed that SEB-specific peripheral tolerance lasted for a long period of time. This did not require boosting of SEB but a single injection was sufficient. Interestingly, CD4+, Vβ8+ T-cell tolerance in SEB-primed euthymic mice was also long lasting [Migita, personal observation]. This seems not to be unique for SEB-induced tolerance but is a common phenomena in the peripheral T-cell tolerance seen in various antigen systems [51, 59–62]. It is not clear why this state is so long-lived. Although no evident mechanism has yet been reported, it is possible that antigen is retained for a long time in a form that induces T-cell unresponsiveness by APC. It is also possible that the molecular change in anergic T cells is so stable that newly synthesized intracellular proteins can not correct it in the short term. Alternatively, a certain cellular mechanism may keep T cells in unresponsiveness. Whereas such a 'suppressor cell' has not clearly been demonstrated to exist, a report showing that γ/δ T cells break oral tolerance [63] may suggest the existence of a suppressor cellular regulatory system in peripheral T-cell tolerance.

Concluding Remarks and Remaining Questions

These studies have been aimed at characterizing two major phenomena in T cells of bacterial superantigen challenged animals: (1) activation induced clonal deletion, and (2) clonal functional unresponsiveness or anergy. As documented above, these changes result in specific T-cell tolerance to the activating bacterial superantigen in vitro. The mechanisms resulting in clonal deletion and anergy are poorly understood. A fundamental question is why some Vβ8+ T cells die and others become anergic. Possibly, affinity differences of Vβ8 domains for the SEB-MHC complex in individual T cells may result in qualitative activation differences and cause either anergy or deletion in T cells. However, this hypothesis seems unlikely because T cells from *scid* mice transgenic for γ,δ TCR genes, in which all cells must have the same affinity, were also partly deleted and partly anergized when strongly stimulated by antigen [Spaner et al., submitted]. Intrinsic differences other than TCR affinity, such as heterogeneity of lymphokine production in different T-cell subsets, was also suggested to affect the fate of T cells [64]. Alternatively, differences of APC which present SEB may be important in determining the fate of activated T cells. Since bacterial superantigens were shown to cause activation in APC by cross-linking class II MHC [65, 66], there may be a veto effect from activated superantigen-presenting cells which alter the fate of interacted T cells. Furthermore, hematopoietic cells were suggested to be more likely to induce tolerance than nonhematopoietic cells. Differences in the nature of APC may then largely affect the fate of activated specific T cells [67].

It is also unclear whether the pathway that exists between anergy and death is linear or parallel. The SEB-primed spleen T cells lost responsiveness to SEB on day 4 when Vβ8+ T-cells deletion was still ongoing. Therefore, it seems to be the case that some Vβ8+ T cells become anergic before they die. However, the CD4+, Vβ8+ T cells which remain in SEB-primed mice after 2 weeks regained responsiveness after four months in vivo. This observation may indicate the existence of a heterogeneity of unresponsiveness. One results in the death of cells probably because of activation exhaustion. Another may be less dependent on proliferation or IL-2R induction and be more like the anergy induced in T-cell clones in vitro.

Interestingly, the results presented here seem to indicate that SEB-priming does not result in clonal expansion or memory of CD4+, Vβ8+ T cells. This is opposite to the current ideas of antigen-induced secondary T-cell responses [68], where antigenic stimulation results in more efficient secondary responses and the scale of memory is proportional to the antigenicity. Just as selection of developing thymocytes requires a different amount of TCR stimulati on (i.e. too high → negative selection, too low → neglect). Perhaps the development of memory has a similar window. SEB stimulation may not result in memory because the superantigen-induced T-cell stimulation is extremely high and may exceed the upper limit which can be processed as antigen-specific T-cell memory.

SEs have emerged as one of the strongest reagents which modulate the immune system in both negative and positive directions. From an immunological and etiological view, data on their in vivo effects will provide insights into the significance of these particular immunological substances and may develop a unique application of these reagents for disease therapy [69].

Acknowledgments

We wish to thank Dr. M. Bevan for provision of monoclonal antibody specific for Vβ8 (F23.1). We also thank C.A. Smith and J. Xu for technical assistance and D. Spaner,

R. Miller and J. Roder and L. Siminovitch for critical reading of the manuscript. This work was supported by grants from the Arthritis Society of Canada and National Cancer Institute of Canada and from the Medical Research Council.

References

1   Bergdoll MS: Enterotoxins; in Easmon CSF, Adlam C (eds): Staphylococci and Staphylococcal Infections, vol 2, chap 16: The Organism in vivo and in vitro. New York, Academic Press, 1983, pp 559–598.
2   Langford MP, Stanton GJ, Johnson HM: Biological effects of staphylococcal enterotoxin A on human peripheral lymphocytes. Infect Immun 1978;22:62–68.
3   Fleischer B, Schrezenmeier H: T cell stimulation by staphylococcal enterotoxins. Clonally variable response and requirement for major histocompatibility complex class II molecules on accessory or target. J Exp Med 1988;167:1697–1707.
4   Janeway CA Jr, Yagi J, Conrad PJ, Katz ME, Jones B, Vroegop S, Buxter S: T cell responses to Mls and to bacterial proteins that mimic its behavior. Immunol Rev 1989;107:61–88.
5   Fraser JD: High-affinity binding of staphylococcal enterotoxins A and B to HLA-DR. Nature 1989;339:221–223.
6   Fischer H, Dohlsten M, Lindvall M, Sjogren H-O, Carlsson R: Binding of staphylococcal enterotoxin A to HLA-DR on B cell lines. J Immunol 1989;142:3151–3157.
7   Mollick JA, Cook RG, Rich RR: Class II MHC molecules are specific receptors for staphylococcal enterotoxin A. Science 1989;244:817–821.
8   White J, Herman A, Pullen AM, Kubo R, Kappler JW, Marrack P: The Vβ-specific superantigen staphylococcal enterotoxin B: Stimulation of mature T cells and clonal deletion in neonatal mice. Cell 1989;56:27–35.
9   Herman A, Kappler JW, Marrack P, Pullen AM: Superantigens: Mechanism of T cell stimulation and role in immune responses. Annu Rev Immunol 1991;9:745–772.
10 Festenstein H: Immunogenetic and biologic aspects of in vitro lymphocyte allotransformation (MLR) in the mouse. Transplant Rev 1973;15:62–88.
11 Abe R, Hodes R: Properties of the Mls system: A revised formulation of Mls genetics and an analysis of T-cell recognition of Mls determinants. Immunol Rev 1989;107: 5–28.
12 Kappler JW, Roehm N, Marrack P: T cell tolerance by clonal elimination in the thymus. Cell 1987;49:273–280.
13 Kappler JW, Staerz U, White J, Marrack PC: Self-tolerance eliminates T cells specific for Mls-modified products of the major histocompatibility complex. Nature 1988;332:35–40.
14 MacDonald HR, Schneider R, Lees RK, Howe RC, Acha-Orbea H, Festenstein H, Zinkernagel RM, Hengartner H: T cell receptor Vβ use predicts reactivity and tolerance to Mlsᵃ-encoded antigens. Nature 1988;332:40–45.
15 Carlsson R, Sjögren H-O: Kinetics of IL-2 and interferon-gamma production, expression of IL-2 receptors, and cell proliferation in human mononuclear cells exposed to staphylococcal enterotoxin A. Cell Immunol 1985;96:175–183.
16 Fisher H, Dohlsten M, Lindvall M, Andersson U, Hedlund G, Ericsson P, Hansson J, Sjogren H-O: Production of TNF-α and TNF-β by staphylococcal enterotoxin A activated human T cells. J Immunol 1990;144:4663–4669.

17 Kawabe Y, Ochi A: Selective anergy of Vβ8⁺, CD4⁺ T cells in staphylococcus enterotoxin B-primed mice. J Exp Med 1990;172:1065–1070.

18 Herrmann T, Maryanski JL, Romero P, Fleischer B, MacDonald HR: Activation of MHC class I-restricted CD8⁺ CTL by microbial T cell mitogens: Dependence upon MHC class II expression of the target cells and Vβ usage of the responder T cells. J Immunol 1990;144:1181–1186.

19 Lando PA, Hedlund G, Dohlsten M, Kalland T: Bacterial superantigens as anti-tumor agents: induction of tumor cytotoxicity in human lymphocytes by staphylococcal enterotoxin A. Cancer Immunol Immunother 1991;33:231–237.

20 Beisel WR: Staphylococcus aureus enterotoxins; in Newberne PM (ed): Trace Substances and Health. New York, Marcel Dekker, 1976, pp 12–27.

21 Bergdoll MS: Enterotoxins; in Montic TC, Kadis S, Ajl SJ (eds): Microbial Toxins. New York, Academic Press, 1970, pp 265–326.

22 Peavy DL, Adler WH, Smith RT: The mitogenic effects of endotoxin and staphylococcal enterotoxin B on mouse spleen cells and human peripheral lymphocytes. J Immunol 1970;105:1453–1458.

23 Zehavi-Willner T, Shenberg E, Barnea A: In vivo effect of staphylococcal enterotoxin A on peripheral blood lymphocytes. Infect Immun 1984;44:401–405.

24 Marrack P, Blackman M, Kushnir E, Kappler J: The toxicity of staphylococcal enterotoxin B in mice is mediated by T cells. J Exp Med 1990;171:455–464.

25 Kawabe Y, Ochi A: Programmed cell death and extrathymic reduction of Vβ8⁺CD4⁺ T cells in mice tolerant to Staphylococcus aureus enterotoxin B. Nature 1991;349: 245–248.

26 MacDonald HR, Baschieri S, Lees RK: Clonal expansion precedes anergy and death of Vβ8⁺ peripheral T cells responding to staphylococcal enterotoxin B in vivo. Eur J Immunol 1991;21:1963–1966.

27 Newell KA, Ellenhorn JD, Bruce DS, Bluestone JA: In vivo T-cell activation by staphylococcal enterotoxin B prevents outgrowth of a malignant tumor. Proc Natl Acad Sci USA 1991;88:1074–1078.

28 Ben-Nun A: Staphylococcal enterotoxin B as a potent suppressant of T cell proliferative responses in rats. Eur J Immunol 1991;21:815–818.

29 Mercep M, Blueston JA, Noguchi PD, Ashwell JD: Inhibition of transformed T cell growth in vitro by monoclonal antibodies directed against distinct activating molecules. J Immunol 1988;140:324–335.

30 Ucker DS, Ashwell JD, Nickas G: Activation-driven T cell death. I. Requirements for de novo transcription and translation and association with genome fragmentation. J Immunol 1989;143:3461–3469.

31 Herman A, Labrecque N, Thibodeau J, Marrack P, Kappler JW, Sekaly R-P: Identification of the staphylococcal enterotoxin A superantigen binding site in the β1 domain of the human histocompatibility antigen HLA-DR. Proc Natl Acad Sci USA 1991;88:9954–9958.

32 Dellabona P, Peccoud J, Kappler J, Marrack P, Benoist C, Mathis D: Superantigens interact with MHC class II molecules outside of the antigen groove. Cell 1990;62: 1115–1121.

33 Gascoigne NRJ, Ames KT: Direct binding of secreted T-cell receptor βchain to superantigen associated with class II major histocompatibility complex protein. Proc Natl Acad Sci USA 1991;88:613–616.

34 Emmrich F: Cross-linking of CD4 and CD8 with the T cell receptor complex:

quaternary complex formation and T cell repertoire selection. Immunol Today 1988; 9:296–299.

35 Collins RJ, Harmon BV, Souvlis T, Pope JH, Kerr JFR: Effects of cycloheximide on B-chronic lymphocytic leukaemic and normal lymphocytes in vitro: Indiction apoptosis. Br J Cancer 1991;64:518–522.

36 Cohen JJ: Programmed cell death in the immune system. Adv Immunol 1991;50:55–85.

37 Ashwell JD, Cunningham RE, Noguchi PD, Hernandez D: Cell growth cycle block of T cell hybridomas upon activation with antigen. J Exp Med 1987;165:173–194.

38 Sambhara SR, Miller RG: Programmed cell death of T cells signaled by the T cell receptor and the α_3 domain of class I MHC. Science 1991;252:1424–1427.

39 Lenardo MJ: Interleukin-2 programs mouse $\alpha\beta$ T lymphocytes for apoptosis. Nature 1991;353:858–861.

40 Zacharchuk CM, Mercep M, Chakraborti PK, Simons SS, Ashwell HD: Programmed T lymphocyte death. Cell activation and steroid-induced pathways are mutually antagonistic. J Immunol 1990;145:4037–4045.

41 Crabtree GR: Contingent genetic regulation events in T lymphocyte activation. Science 1989;243:355–361.

42 Cohen JJ, Duke RC, Sellins KS: Stimulation by superantigen. Nature 1991;352:199–200.

43 Ochi A, Kawabe Y: Death by superantigen. Nature 1992;355:211–212.

44 Duke RC, Cohen JJ: IL-2 addiction: withdraw of growth factor activates a suicide program in dependent cells. Lymphokine Res 1986;5:289–299.

45 Schwartz RH: A cell culture model for T lymphocyte clonal anergy. Science 1990; 248:1349–1356.

46 Jenkins MK, Schwartz RH: Antigen presentation by chemically modified splenocytes induces antigen-specific T cell unresponsiveness in vitro and in vivo. J Exp Med 1987;165:302–319.

47 Quill H, Schwartz HR: Stimulation of normal inducer T cell clones with antigen presented by purified Ia molecules in planar lipid membranes: specific induction of a long-lived state of proliferative nonresponsiveness. J Immunol 1987;138:3704–3712.

48 Jenkins MK, Ashwell JD, Schwartz RH: Allogeneic non-T spleen cells restore the responsiveness of normal T cell clones stimulated with antigen and chemically modified antigen-presenting cells. J Immunol 1988;140:3324–3330.

49 Jenkins MK, Chen C, Jung G, Mueller D, Schwartz RH: Inhibition of antigen-specific proliferation of type 1 murine T cell clones after stimulation with immobilized anti-CD3 monoclonal antibody. J Immunol 1990;144:16–22.

50 Mueller DL, Jenkins MK, Schwartz RH: Clonal expansion versus functional clonal inactivation: A costimulatory signalling pathway determines the outcome of T cell antigen receptor occupancy. Annw Rev Immunol 1989;7:445–480.

51 Rammensee H, Kroschewski R, Frangoulis B: Clonal anergy induced in mature $V_\beta b^+$ T lymphocytes immunized with Mls-1^a antigen. Nature 1989;339:541–544.

52 Webb S, Morris C, Sprent J: Extrathymic tolerance of mature T cells. Clonal elimination as a consequence of immunity. Cell 1990;63:1249–1256.

53 Rellahan BL, Jones LA, Kruisbeek AM, Fry AM, Matis LA: In vivo induction of anergy in peripheral $V\beta8^+$ T cells by Staphylococcal enterotoxin B. J Exp Med 1990; 172:1091–1100.

54 Blackman MA, Gerhard-Burgert H, Woodland DL, Palmer E, Kappler JW, Marrack P: A role for clonal inactivation in T cell tolerance to Mls-1[a]. Nature 1990;345:540–542.

55 Janeway CA Jr: Selective elements for the Vβ region of the T cell receptor: Mls and the bacterial toxic mitogens. Adv Immunol 1991;50:1–53.

56 O'Rourke AM, Mescher MF, Webb SR: Activation of polyphosphoinositide hydrolysis in T cells by H-2 allo antigen but not Mls determinants. Science 1990;249:171–174.

57 Liu H, Lampe MA, Iregui MV, Cantor H: Conventional antigen and superantigen may be coupled to distinct and cooperative T-cell activation pathways. Proc Natl Acad Sci USA 1991;88:8705–8709.

58 Blackman MA, Finkel TH, Kappler JW, Marrack P: Altered antigen receptor signaling in anergic T cells from self-tolerant T-cell receptor β-chain transgenic mice. Proc Natl Acad Sci USA 1991;88:6682–6686.

59 Miller RG, Phillips RA: Reduction of the in vitro cytotoxic lymphocyte response produced by in vivo exposure to semiallogeneic cells. Recruitment or active suppression? J Immunol 1976;117:1913–1921.

60 Martin RD, Miller GR: In vivo administration of histocompatible lymphocytes leads to rapid functional deletion of cytotoxic T lymphocyte precursors. J Exp Med 1989;170:679–690.

61 Roser BJ: Cellular mechanisms in neonatal and adult tolerance. Immunol Rev 1989;107:179–202.

62 Macphail S, Stutman O: Specific neonatal induction of functional tolerance to allogeneic Mls determinants occurs intrathymically and the tolerant status is Mls haplotype specific. J Immunol 1989;143:1795–1800.

63 Fujihashi K, Kiyono H, Aicher WK, Green DR, Singh B, Eldrige JH, McGhee JR: Immunoregulatory function of CD3[+], CD4[-], and CD8[-] T cells: γδ T cell receptor-positive T cells from nude mice abrogate oral tolerance. J Immunol 1989;143:3415–3422.

64 Bandeira A, Mengel J, Burlen-Defranoux G, Coutinho A: Proliferative T cell anergy to Mls-1[a] does not correlate with in vivo tolerance. Int Immunol 1991;3:923–932.

65 Spertini F, Spits H, Geha RS: Staphylococcal enterotoxins deliver activation signals to human T-cell clones via major histocompatibility complex class II molecules. Proc Natl Acad Sci USA 1991;88:7533–7537.

66 Mourad W, Scholl P, Diaz A, Geha R, Chatila T: The staphylococcal toxic syndrome toxin 1 triggers B cell proliferation and differentiation via major histocompatibility complex-unrestricted cognate T/B cell interaction. J Exp Med 1989;170:2011–2022.

67 Gao Er-Kai, Kosaka H, Surh CD, Sprent J: T cell contact with Ia antigens on nonhemopoietic cells in vivo can lead to immunity rather than tolerance. J Exp Med 1991;174:435–446.

68 Cerottini JC, Macdonald HR: The cellular basis of T cell memory. Annu Rev Immunol 1989;7:77–89.

69 Kim C, Siminovitch KA, Ochi A: Reduction of lupus nephritis in MLR/lpr mice by a bacterial superantigen treatment. J Exp Med 1991;174:1431–1437.

Atsuo Ochi, PhD, Division of Neurobiology and Molecular Immunology,
Samuel Lunenfeld Research Institute, University of Toronto, Mount Sinai Hospital,
600 University Avenue, Toronto ON M5G 1X5 (Canada)

Fleischer B (ed): Biological Significance of Superantigens.
Chem Immunol. Basel, Karger, 1992, vol 55, pp 137–145

Mycoplasma arthritidis-Derived Superantigen

Lothar Rink, Holger Kirchner

Institute of Immunology and Transfusion Medicine,
University of Lübeck Medical School, Lübeck, FRG

Introduction

Mycoplasms are members of the unique taxon mollicutes, which defines the class of cell wall free procaryotes. Mycoplasms are the smallest autoreplicative microorganisms with a metabolism of their own [1]. The genome size of mycoplasms is in the order of 5×10^8 Da [2]. For this reason, mycoplasms have a limited synthetic capability and live parasitically or saprophytically.

Mycoplasma arthritidis is a pathogen for rodents and induces an acute inflammatory infection in mice and rats. In mice this inflammation is followed by a chronic joint disease [3]. Pathogenic effects of *M. arthritidis* in human diseases, especially in the pathogenesis of rheumatoid arthritis are still the subject of discussion.

In 1982 Cole et al. [4] were the first to show that the cell-free supernatant of *M. arthritidis* has immunomodulatory effects. Somewhat later, our group [5] also found the T-cell stimulatory effect of *M. arthritidis*-derived superantigen (MAS). In the following years these two groups studied the principle of MAS. We showed first [6] that T-cell activation is not antigen specific. Later, the group of Fleischer, in cooperation with us, postulated that MAS is a protein comparable to the staphylococcal enterotoxins in their T-cell stimulatory property [7]. Since Marrack and Kappler [8] included MAS in the group of staphylococcus enterotoxin-like toxins, MAS is now considered to be a superantigen. Excluding the Mls antigens, MAS is the only member of this group which is not produced by gram-positive bacteria.

MAS is exclusively produced by *M. arthritidis*. No other mycoplasm tested produced a comparable soluble T-cell mitogenic protein [9]. Genetic comparisons showed that mycoplasms are phylogenetically related to the gram-positive clostridia [10]. If there is any sequence homology between MAS and the other superantigens, the superantigen principle must be as old as the phylogenetic divergence of these taxons.

However, MAS has not yet been sequenced. Purification of MAS is very difficult, since there are only very small amounts of MAS in the mycoplasma culture. On the other hand, *M. arthritidis* has to be cultured in a complete broth containing yeast extract and horse serum. Additionally, MAS has a high affinity to glass and plastic surfaces as well as to high-molecular-weight proteins of the mycoplasmal broth [11], especially serum albumin [12]. For this reason all studies about MAS and its immunomodulatory effect were done with the unpurified cell-free supernatant of *M. arthritidis* or with partially purified preparations of the supernatant.

Despite these problems, some biochemical data of MAS were established. MAS has a molecular weight of 15 kDa, as detected by gel filtration [11], or 27–30 kDa, as shown by SDS-PAGE [11, 12]. MAS is a hydrophobic protein [13] with an isoelectric point around pH 9 [11]. The protein is heat-labile at 56 °C [11] and labile to serine proteases [unpubl. data of our group].

MAS and the MHC Molecules

As other superantigens, MAS does not need processing [7, 14]. This mechanism was shown for MAS before the group of superantigens was defined. The reaction of T-cells to MAS is strictly dependent on the presentation of MAS by MHC class II-bearing accessory cells (AC) [4, 15, 16]. However, not all MHC class II molecules can present MAS to the T-cell receptor (TCR). These investigations were made by Cole et al. [17] in different systems. First it was shown that MAS only reacts with some mouse strains. Mouse strains with the H2 haplotypes a, d, j, k, p, r, u and v react with proliferation on stimulation with MAS, whereas mouse strains with the H2 haplotypes b, f, q and s fail to respond to MAS. Further investigations showed that the whole H2 complex is not needed. MAS interacts only with the Eα-chain of the MHC II heterodimer [4, 14, 18]. All mouse strains which respond to MAS express an intact Eα-molecule on the cell surface of AC. It was shown that the response is independent of the Eβ-chain, because combinatorial EαAβ molecules are also able to present MAS [19]. This observation was confirmed in different experimental systems. Antibodies against Eα blocked the T-cell proliferation mediated by MAS [19]. Eα-transfected fibroblasts were able to present MAS, whereas other transfected fibroblasts which lack in Eα failed to respond [19]. Finally, Eα transgenic mice of a normally MAS nonresponder mouse strain reacted to MAS [19]. All these systems were based on living AC with a functional Eα molecule. However, MAS could be presented by fixed AC [14] as well as by liposomes coated onto glass beads with incorporated I-E molecules from solubilized AC [20]. Consequently, there is no doubt about the major role of Eα in the response of murine leukocytes to MAS. This specificity for Eα shows, in contrast to the other superantigens, that MAS is not only specific for some Vβ regions of the TCR. *Thus, MAS has a specificity for both sites of the superantigen bridge.*

The analogous molecule to the murine I-Eα in the human system seems to be the α-chain of HLA-DR. MAS also stimulates human peripheral blood mononuclear cells (PBM) for proliferation. This proliferation could be blocked by antibodies to the nonpolymorphic regions of the HLA-DR α-chain [7, 21].

However, we showed that MAS stimulated macrophages without cross-linking the MHC II complex and the TCR. MAS induced interleukin-6 (IL-6) [13] and tumor necrosis factor-α (TNF-α) [unpubl. data of our group] in murine bone marrow-derived macrophages (BMM) of MAS responder mouse strains. The cultures of BMM were of 95% purity and a presence of T-cells could be ruled out. The response was specific for MAS, since BMM from C3H/HeJ mice, which fail to respond to lipopolysaccharide (LPS) showed high levels of these cytokines after induction with MAS. On the other hand, BMM from C57Bl/6 mice which fail to respond to MAS and respond to LPS showed no induction of IL-6 and TNF-α by MAS. Thus, artefacts by contaminating LPS could be ruled out. Therefore, the direct induction of these monokines seems to be also dependent on I-Eα, but independent of cross-linking the MHC II complex and the TCR.

The murine macrophage cell line J774.1 showed a tumoricidal effect on 3T12-3 mouse embryo cells after stimulation with MAS [22]. It seems possible to us that this effect was caused by induction of TNF-α.

Stimulation of Lymphocytes by MAS

Leukocyte cultures from man, mouse and rat react to stimulation with MAS, whereas leukocytes from rabbit, guinea pig, sheep and cow fail to respond [23]. It is not clear if the non-responsiveness is caused by an absent binding site on the MHC complex or missing TCR binding regions.

The activation of T cells is dependent on AC, but is not restricted on self MHC presentation. MAS could stimulate T cells when it is presented by an MHC class II molecule with the appropriate binding site and it is irrelevant if the MHC is syngenic, allogenic or xenogenic [17, 24]. For example, cells transfected with murine I-E can present MAS to human T cells [7]. However, not all T-cells are activated by MAS. Furthermore, the antigen specificity of the responsive T cells does not play a role [7, 14, 16, 25].

Both CD4+ and CD8+ human T-cell clones respond to MAS [7]. Even both types of TCR, α/β as well as γ/δ, are responsive to MAS [7]. Also, in the murine system the CD4 molecule is not necessary for T-cell activation by MAS [20]. By using the RIIIS mouse strain it was first observed that splenocyte cultures from an I-Eα^+ mouse strain did not proliferate after MAS stimulation [26]. Cross-breeding experiments showed that the nonrespon-siveness of RIIIS mice followed Mendel's law. The fact that both parents were I-Eα^+ underlined that the nonresponsiveness is located on gene loci other than MHC. Antibodies against the $V\beta$ regions of the α/β TCR clarified that the unresponsiveness is based on the TCR $V\beta$ genes [27]. In flow cytometry studies it could be shown that T cells expressing the $V\beta$ regions 6, 8.1, 8.2, and 8.3 proliferate after stimulation with MAS [28]. There is a preference in activation of $V\beta$ 8.2 expressing T cells. However, it seems that MAS has a lower affinity to a TCR molecule different from the $V\beta$ regions 6 and 8, because there is a background proliferation of lymphocytes that fail to express any of these $V\beta$ regions [28].

MAS and Mls-1^a use the same $V\beta$ regions 6 and 8.1 [8]. For this reason the unresponsiveness of some mouse strains to MAS could be caused by the T-cell depletion of these $V\beta$ regions by the Mls antigens.

T-cell proliferation was the first effect that was seen after stimulation of mixed lymphocyte cultures with MAS. However, MAS induced various cytokines in mixed lymphocyte cultures. First it was shown that MAS induced interferon-γ (IFN-γ) in human [21] and murine [5] lymphocytes. Recently, we showed that MAS induced IL-1α, IL-1β, IL-2, IL-4, IL-6 and TNF-α in human PBM [29]. The induction of the T-cell products was dependent on the presence of AC, whereas the induction of monokines also took place in separated monocytes. In murine splenocyte cultures, we could also detect IL-6 [13] and TNF-α [unpubl. data]. In the human leukocyte cultures, we observed a dissociation between the secretion of the cytokines IL-4 and IFN-γ. The differences were based on the source of the leukocytes [29]. We also measured differences between the induction of cytokines in murine and human leukocytes. Furthermore, we showed that MAS is less active in induction of these cytokines in PBM in comparison to the staphylo-coccal enterotoxins [unpubl. data]. These findings correspond to the previous studies that human PBM proliferate to a higher degree after stimulation with SE in comparison to MAS [30]. On the other hand, murine splenocytes proliferate better after stimulation with MAS in comparison to SE [30]. These studies showed that the SE has a higher specificity for human cells, whereas MAS is more specific for murine cells. Therefore, in our opinion it seems possible that superantigens are more specific for cells of their natural host.

The reaction of T cells is strictly dependent on AC as described above. However, there is some evidence that the effect of MHC class II could be partially substituted. Nylon wool-purified T cells did not normally proliferate after induction with MAS. After treatment of those purified T cells with IL-1, the proliferation rate was partially reconstituted [31]. This shows that the induction of IL-1 is possibly of relevance as a second signal. Another indication was that splenocytes from C57Bl/6 mice (Eα⁻), which are normally MAS nonresponder, proliferated after treatment with 2-mercaptoethanol [32]. Furthermore, MAS activated T cells directly when added in combina-tion with the protein kinase C activator phorbol myristate acetate (PMA) [7]. MAS stimulation also increases the intracellular Ca^{2+} level in the Jurkat T-cell line [7]. Since this cell line is MHC class II negative, MAS must also interact separately with the TCR without cross-linking with the MHC [7]. However, for T-cell proliferation there must be a second signal because the Jurkat line did not proliferate after MAS stimulation [7]. This second signal could be protein kinase C activation via PMA or through signal transduction of the IL-1 or IL-6 receptors. The fact that T cells proliferate after stimulation with MAS presented by Eα-coated liposome glass beads [20], suggests that the superantigen bridge could be a second signal, too.

Polyclonal B-cell activation and immunoglobulin secretion of human [33] and murine [34] B cells were observed in cocultures of B cells with MAS-responsive T_H cells and MAS. As described above, the superantigen bridge of MAS must react as a signal for B cells. Another B-cell-activating pathway could be the induction of the B-cell-stimulating factors IL-4 and IL-6 as described before.

MAS and Diseases

M. arthritidis induces an acute inflammatory infection in mice and rats, followed by a chronic joint disease in mice [3]. The infection and the resulting experimentally induced arthritis is accompanied by a toxic shock syndrome. The toxic shock syndrome after intravenous injection possibly resulted from the MAS-induced inflammatory cytokines (IL-6 and TNF-α) as described earlier. The fact that only MAS-responsive I-E$^+$ mouse strains get the shock syndrome [35] underlined the hypothesis that this toxic shock is mediated by MAS. Furthermore, only MAS responder mouse strains suffer from necrosis at the subcutaneous injection site of *Mycoplasma arthritidis* [36].

In all studies MAS showed immunosuppressive effects after intravenous injection [37] in MAS-responsive mouse strains. Splenocytes of these mice showed a decreased proliferation after stimulation with MAS [37]. This suppressive effect may be caused by T-cell depletion as described for other superantigens. However, MAS did not induce arthritis in mice but intra-articular injection of MAS into rats induced a synovitis [38]. These may also be effects of the induced inflammatory and necrotic cytokines. The fact that these inflammatory cytokines are also detected in the synovial fluid of patients with rheumatoid arthritis [39] may also represent a hint to the pathogenic capability of these cytokines.

Conclusions and Perspectives

MAS induces monocytes/macrophages to produce IL-1, IL-6 and TNF-α. These cytokines may play a role in inflammation, necrosis and in B-cell as well as T-cell stimulation. Furthermore, the induced IL-2, IL-4 and IFN-γ may be activators for these cell types. The cytokines, as secondary signals after the cell contact through the superantigen bridge, may induce the cells involved in different ways. The activation of autoreactive T cells, B-cell-stimulating T$_H$ cells and autoantibody-producing B cells might result in autoimmune disease. Especially for B-cell activation, there are some observations that underline the role of superantigens in autoimmune diseases [40].

References

1   Maniloff J, Morowitz HJ: Cell biology of mycoplasmas. Bacteriol Rev 1972;36:263–290.

2   Bak AL, Black FT, Christiansen C, Freundt EA: Genome size of mycoplasmal DNA. Nature 1969;224:1209–1210.

3   Cole BC, Ward JR, Jones RS, Cahill JF: Chronic proliferative arthritis of mice induced by *Mycoplasma arthritidis*. Infect Immun 1971;4:344–355.

4   Cole BC, Sullivan GJ, Daynes RA, Sayed IA, Ward JR: Stimulation of mouse lymphocytes by a mitogen derived from *Mycoplasma arthritidis*. II. Cellular requirements for T-cell transformation mediated by a soluble Mycoplasma mitogen. J Immunol 1982;128:2013–2018.

5   Kirchner H, Nicklas W, Giebler D, Keyssner K, Berger R, Storch E: Induction of interferon gamma in mouse spleen cells by culture supernatant of *Mycoplasma arthritidis*. J Interferon Res 1984;4:389–397.

6   Kirchner H, Bauer A, Moritz T, Herbst F: Lymphocyte activation and induction of interferon gamma in human leucocyte cultures by the mitogen in *Mycoplasma arthritidis* supernatant (MAS). Scand J Immunol 1986;24:609–613.

7   Matthes M, Schrezenmeier H, Homfeld J, Fleischer S, Malissen B, Kirchner H, Fleischer B: Clonal analysis of human T-cell activation by the *Mycoplasma arthritidis* mitogen (MAS). Eur J Immunol 1988;18:1733–1737.

8   Marrack P, Kappler J: The staphylococcal enterotoxins and their relatives. Science 1990;248:705–711.

9   Weisburg WG, Tully JG, Rose DL, Petzel JP, Oyaizu H, Yang D, Mandelco L, Sechrest J, Lawrence TG, Van Etten J, et al: A phylogenetic analysis of the mycoplasmas: Basis for their classification. J Bacteriol 1989;171:6455–6467.

10   Woese CR, Maniloff J, Zablen LB: Phylogenetic analysis of the mycoplasmas. Proc Natl Acad Sci USA 1980;77:494–498.

11   Atkin CL, Cole BC, Sullivan GJ, Washburn LR, Wiley BB: Stimulation of mouse lymphocytes by a mitogen derived from *Mycoplasma arthritidis*. J Immunol 1986;137:1581–1589.

12   Kirchner H, Brehm G, Nicklas W, Beck R, Herbst F: Biochemical characterization of the T-cell mitogen derived from *Mycoplasma arthritidis*. Scand J Immunol 1986;24:245–249.

13   Homfeld J, Homfeld A, Nicklas W, Rink L, Weyland A, Kirchner H: Induction of interleukin-6 in murine bone marrow-derived macrophages stimulated by the *Mycoplasma arthritidis* mitogen MAS. Autoimmunity 1990;7:317–327.

14   Cole BC, Aranco BA, Sullivan GJ: Stimulation of mouse lymphocytes by a mitogen derived from *Mycoplasma arthritidis*. IV. Murine T hybridoma cells exhibit differential accessory cell requirements for activation by *M. arthritidis* T-cell mitogen, concanavalin A, or hen egg-white lysozyme. J Immunol 1986;136:3572–3578.

15   Cole BC, Daynes RA, Ward JR: Stimulation of mouse lymphocytes by a mitogen derived from *Mycoplasma arthritidis*. III. Ir gene control of lymphocyte transformation correlates with binding of the mitogen to specific Ia-bearing cells. J Immunol 1982;129:1352–1359.

16   Lynch DH, Gurish MF, Cole BC, Daynes RA: T-cell proliferative responses to a mitogen derived from *Mycoplasma arthritidis* are controlled by the accessory cell. J Immunol 1983;131:1702–1706.

17   Cole BC, Daynes RA, Ward JR: Stimulation of mouse lymphocytes by a mitogen derived from *Mycoplasma arthritidis*. I. Transformation is associated with an H-2-linked gene that maps to the I-E/I-C subregion. J Immunol 1981;127:1931–1936.

18   Cole BC, Griffiths MM, Sullivan GJ, Ward JR: Role of non-RT1 genes in the response of rat lymphocytes to *Mycoplasma arthritidis* T-cell mitogen, concanavalin A and phytohemagglutinin. J Immunol 1986;136:2364–2369.

19 Cole BC, David CS, Lynch DH, Kartchner DR: The use of transfected fibroblasts and transgenic mice establishes that stimulation of T-cells by the *Mycoplasma arthritidis* mitogen is mediated by E alpha. J Immunol 1990;144:420–424.

20 Bekoff MC, Cole BC, Grey HM: Studies on the mechanism of stimulation of T-cells by the *Mycoplasma arthritidis*-derived mitogen. Role of class II IE molecules. J Immunol 1987;139:3189–3194.

21 Cole BC, Thorpe RN: Induction of human gamma interferons by a mitogen derived form *Mycoplasma arthritidis* and by phytohemagglutinin: differential inhibition with monoclonal anti-HLA.DR antibodies. J Immunol 1983;131:2392–2396.

22 Dietz JN, Cole BC: Direct activation of the J774.1 murine macrophage cell line by *Mycoplasma arthritidis*. Infect Immun 1982;37:811–819.

23 Cole BC, Washburn LR, Sullivan GJ, Ward JR: Specificity of a mycoplasma mitogen for lymphocytes from human and various animal hosts. Infect Immun 1982;36: 662–666.

24 Lynch DH, Cole BC, Bluestone JA, Hodes RJ: Cross-reactive recognition by antigen-specific, major histocompatibility complex-restricted T cells of a mitogen derived from *Mycoplasma arthritidis* is clonally expressed and I-E restricted. Eur J Immunol 1986; 16:747–751.

25 Yowell RL, Cole BC, Daynes RA: Utilization of T-cell hybridomas to establish that a soluble factor derived from *Mycoplasma arthritidis* is truly a genetically restricted polyclonal T-cell activator. J Immunol 1983;131:543–545.

26 Cole BC, Tuller JW, Sullivan GJ: Stimulation of mouse lymphocytes by a mitogen derived from *Mycoplasma arthritidis*. VI. Detection of a non-MHC gene(s) in the E alpha-bearing RIIIS mouse strain that is associated with a specific lack of T-cell responses to the *M. arthritidis* soluble mitogen. J Immunol 1987;139:927–935.

27 Cole BC, Kartchner DR, Wells DJ: Stimulation of mouse lymphocytes by a mitogen derived from *Mycoplasma arthritidis*. VII. Responsiveness is associated with expression of a product(s) of the V beta 8 gene family present on the T-cell receptor alpha/beta for antigen. J Immunol 1989;142:4131–4137.

28 Cole BC, Kartchner DR, Wells DJ: Stimulation of mouse lymphocytes by a mitogen derived from *Mycoplasma arthritidis* (MAM). VIII. Selective activation of T-cells expressing distinct V beta T-cell receptors from various strains of mice by the 'superantigen' MAM. J Immunol 1990;144:425–431.

29 Rink L, Kruse A, Nicklas W, Hoyer J, Kirchner H: Induction of cytokines in human peripheral blood and spleen cells by the *Mycoplasma arthritidis*-derived superantigen. Lymphokine Res 1992;in press.

30 Fleischer B, Mittrucker HW, Metzroth B, Braun M, Hartwig U: Mitogenic toxins as MHC class II-dependent probes for T-cell antigen receptors. Behring Inst Mitt 1991; 88:170–176.

31 Bauer A, Giese M, Kirchner H: Role of interleukin 1 in mycoplasma mitogen-induced proliferation of human T-cells. Immunobiology 1989;179:124–130.

32 Moritz T, Giebler D, Gunther E, Nicklas W, Kirchner H: Lymphoproliferative responses of spleen cells of inbred rat strains to *Mycoplasma arthritidis* mitogen. Scand J Immunol 1984;20:365–369.

33 Tumang JR, Posnett DN, Cole BC, Crow MK, Friedmann SM: Helper T-cell-dependent human B-cell differentiation mediated by a mycoplasmal superantigen bridge. J Exp Med 1990;171:2153–2158.

34 Tumang JR, Cherniack EP, Gietl DM, Cole BC, Russo C, Crow MK, Friedman SM: T

helper cell-dependent, microbial superantigen-induced murine B-cell activation: polyclonal and antigen-specific antibody responses. J Immunol 1991;147:432–438.

35 Cole BC, Thorpe RN, Hassel LA, Washburn LR, Ward JR: Toxicity but not arthritogenicity of *Mycoplasma arthritidis* for mice associates with the haplotype expressed at the major histocompatibility complex. Infect Immun 1983;41:1010–1015.

36 Cole BC, Piepkorn MW, Wright EC: Influence of genes of the major histocompatibility complex on ulcerative dermal necropsis induced in mice by *Mycoplasma arthritidis*. J Invest Dermatol 1985;85:357–361.

37 Cole BC, Wells DJ: Immunosuppressive properties of *Mycoplasma arthritidis* T-cell mitogen in vivo: inhibition of proliferative responses to T-cell mitogens. Infect Immun 1990;58:228–236.

38 Cannon GW, Cole BC, Ward JR, Smith JL, Eichwald EJ: Arthritogenic effects of *Mycoplasma arthritidis* T-cell mitogen in rats. J Rheumatol 1988;15:735–741.

39 Bhardwaj N, Santhanam U, Lau LL, Tatter SB, Ghrayeb J, Rivelis M, Steinman RM, Sehgal PB, May LT: IL-6/IFN-beta 2 in synovial effusions of patients with rheumatoid arthritis and other arthritides. Identification of several isoforms and studies of cellular sources. J Immunol 1989;143:2153–2159.

40 Friedman SM, Posnett DN, Tumang JR, Cole BC, Crow MK: A potential role of microbial superantigens in the pathogenesis of systemic autoimmune disease. Arthritis Rheum 1991;34:468–480.

Lothar Rink, Institute of Immunology and Transfusion Medicine, University of Lübeck Medical School, Ratzeburger Allee 160, D-W–2400 Lübeck 1 (FRG)

Fleischer B (ed): Biological Significance of Superantigens.
Chem Immunol. Basel, Karger, 1992, vol 55, pp 146–171

Cellular and Molecular Mechanisms of Immune Activation by Microbial Superantigens: Studies Using Toxic Shock Syndrome Toxin-1

Talal Chatila, Paul Scholl, Narayanaswamy Ramesh, Nikolaus Trede,
Ramsay Fuleihan, Tomohiro Morio, Raif S. Geha

Division of Immunology, The Children's Hospital, and
Department of Pediatrics, Harvard Medical School, Boston, Mass., USA

Introduction

The term superantigens has been coined in reference to a group of molecules which, when bound to MHC class II molecules, stimulate T cells bearing particular T-cell receptor Vβ elements. Superantigens differ in several important aspects, to be discussed later, from classical peptide antigens in their interaction with both MHC class II molecules and with the T-cell receptor. Both endogenous (self) and foreign superantigens have been described. The former are exemplified by the murine minor lymphocyte stimulating determinants, or MLs, recently determined to be retroviral gene products. Foreign superantigens are typified by the staphylococcal enterotoxins, causative agents of food poisoning and toxic shock syndrome (TSS). Our studies have focused on toxic shock syndrome toxin-1 (TSST-1), a major causative agent of TSS, using this staphylococcal exotoxin to probe the effects superantigens exert on the immune system. The following is a review of the superantigenic properties of TSST-1 and a presentation of some recent insights into the cellular and molecular mechanisms of action of this agent.

Role of TSST-1 and Related Exotoxins in Toxic Shock Syndrome

TSS is a severe multisystem disorder characterized by high fever, hypotension, generalized erythroderma, desquamation of the skin and dysfunction of multiple organ systems [Chesney, 1989; Davis et al., 1980; Todd et al., 1978]. TSS is consistently associated with infection with toxigenic strains of *Staphylococcus aureus*, most commonly in the setting of tampon use during menses or following surgery or trauma. Exotoxins secreted by staphylococcal isolates from patients with TSS, most notably toxic shock syndrome toxin-1 (TSST-1) but also the structurally related staphylococcal enterotoxins, play a key role in the pathophysiology of this disease. Unlike randomly tested staphylococcal isolates, the overwhelming majority of staphylococcal isolates from patients with TSS are toxigenic [Bergdoll et al., 1981; Bonventre et al., 1989; Crass and Bergdoll, 1986; Schlievert et al., 1981]. Almost all staphylococcal isolates from menstruation-associated TSS cases and half of the isolates from nonmenstrual TSS cases express TSST-1 alone or, more commonly, in combination with one or more of the staphylococcal enterotoxins [Crass and Bergdoll, 1986]. The majority of TSST-1 negative isolates express one or more of the staphylococcal enterotoxins, particularly staphylococcal enterotoxin B (SEB) [Crass and Bergdoll, 1986; Parsonnet et al., 1986; Schlievert, 1986]. The critical role played by TSST-1 in the genesis of TSS has been established by studies demonstrating that TSST-1-positive strains of *S. aureus* produce a TSS-like illness in rabbits while isogenic, TSST-1 negative strains fail to do so [de Azavedo et al., 1985; Rasheed et al., 1985]. Passive immunization with a neutralizing anti-TSST-1 monoclonal antibody protects rabbits from otherwise lethal infection with TSST-1 producing strains of staphylococci [Best et al., 1988]. Further evidence incriminating TSST-1 in the pathogenesis of TSS came from the demonstration that constant infusion of purified TSST-1 in rabbits also produces a TSS-like illness [Parsonnet et al., 1987] and that rabbits can be protected against this illness by passive immunization with the same neutralizing anti-TSST-1 monoclonal antibody [Bonventre et al., 1988]. Interestingly, infusion of staphylococcal enterotoxins induces a TSS-like illness in rabbits indistinguishable from that following TSST-1 infusion [Parsonnet et al., 1987]. This supported a role for staphylococcal enterotoxins in the pathogenesis of TSS, especially in cases associated with enterotoxin producing but TSST-1 nonproducing strains of staphylococci.

The susceptibility to TSS seems to correlate with poor antibody response to TSST-1 and to enterotoxins [Crass and Bergdoll, 1986]. Thus, while the majority of healthy individuals have protective titers of antibodies against TSST-1, more than 80% of patients with TSS display no detectable levels of anti-TSST-1 antibodies when examined during the acute phase of the disease. Furthermore, most TSS patients fail to develop protective titers of anti-TSST-1 antibodies following convalescence. A similar trend was observed when examining titers of anti-enterotoxin antibodies. This suggested that most patients with TSS may either have an immunodeficiency that impairs their ability to mount a humoral immune response to TSST-1 and staphylococcal enterotoxins or that the toxins inhibit the generation of an antigen-specific antibody response (see below).

Activation of the Immune System by TSST-1

TSST-1 does not exert direct toxic effects on the vast majority of tissues tested, even at very high concentrations [Parsonnet, 1989]. Rather, induction of TSS by TSST-1 may result from massive and unregulated stimulation of the immune system. In vitro tests reveal TSST-1 to be a powerful activator of lymphocytes and monocytes. TSST-1 is a potent inducer of IL-1 [Ikejima et al., 1984; Parsonnet et al., 1986; Parsonnet et al., 1985] and TNF [Fast et al., 1989; Jupin et al., 1988; Parsonnet and Gillis, 1988] production in human monocytes, more potent on a molar basis than lipopolysaccharides (LPS). TSST-1 is also a T-cell mitogen [Calvano et al., 1984], and it induces the production of myriad T-cell lymphokines including interleukin-2 (IL-2) [Chatila et al., 1988; Micusan et al., 1986; Uchiyama et al., 1986], interferon-γ, lymphotoxin (TNFβ) [Jupin et al., 1988] and colony-stimulating factors [Galelli et al., 1989]. Administration of large quantities of monokines, lymphokines or polyclonal activators of T lymphocytes to experimental animals or human subjects results in a shock state analogous to that observed with TSST-1. For example, infusion of IL-1 and TNF into rabbits reproduces the hemodynamic effects of TSST-1 infusion [Ikejima et al., 1989]. Also, the toxicity of IL-2 administered in large quantities for treating human subjects with cancer [Rosenberg et al., 1988] and of monoclonal antibodies directed against the T-cell receptor associated CD3 complex and administered to transplant recipients [Chatenoud et al., 1990] is similar in many respects to the disturbances observed in patients suffering from TSS. These studies raised the possibility that the toxicity of TSST-1 is intimately related to its potent activation of immune cells.

The notion that the immunostimulatory effects of TSST-1 play an important role in the pathogenesis of TSST-1 is supported by evidence of severe histopathologic changes in the lymphoid tissues of fatal cases of TSS [Paris et al., 1982]. Lymph nodes display marked histiocytosis and hemo-phagocytosis in the interfollicular areas. This is frequently associated with intrafollicular or follicular lymphoid hyperplasia. Lymphoid hyperplasia is also manifest in the gastrointestinal tract, while the spleen demonstrates sinus histiocytosis and lymphoid depletion. In addition, lymphocytic infiltrates are manifest in the periportal areas of the liver and around small vessels in the skin. These findings are mirrored in rabbits inoculated with toxigenic strains of *Staphylococcus aureus* or given a constant infusion of purified TSST-1 [de Azavedo et al., 1985; Parsonnet et al., 1987; Rasheed et al., 1985].

Direct proof for an obligatory role for immune cell activation in the pathogenesis of TSS has been provided by studies using immunosuppressive agents. In one study, concurrent administration of steroids together with purified TSST-1 aborted the emergence of a TSS-like illness and prevented death in four out of four rabbits tested [Parsonnet et al., 1987]. In contrast, four of four rabbits receiving TSST-1 and saline died of a TSS-like illness. In another study, a murine model was developed to study the toxicity of another exotoxin implicated in TSS, SEB [Marrack et al., 1990]. It was demonstrated that inhibition of T lymphocyte activation with cyclosporine greatly attenuated the morbidity and mortality resulting from SEB injection in these mice.

Our understanding of the mechanism of action of TSST-1 in stimulating the immune system has been greatly clarified with the discovery that Ia molecules serve as receptors for TSST-1 [Scholl et al., 1989a; Uchiyama et al., 1989] and that Ia-bound TSST-1 behaves as a superantigen that can stimulate large numbers of T cells in a Vβ2 specific manner [Choi et al., 1989]. TSST-1 can also act via Ia molecules to deliver growth promoting signals to lymphocytes and to induce the production of IL-1 and TNF in monocytes. The binding of TSST-1 to Ia molecules is a property that is shared by all TSS-associated staphylococcal enterotoxins [Fischer et al., 1989; Fraser, 1989; Mollick et al., 1989]. These toxins also behave as superantigens that activate large populations of T lymphocytes bearing distinct Vβ products [Choi et al., 1989; Kappler et al., 1989]. Remarkably, streptococcal pyrogenic toxin A, which is related to the staphylococcal exotoxins and which has been implicated in causing a TSS-like illness [Cone et al., 1987], has also been demonstrated to bind to Ia molecules and to act as a superantigen [Imanishi et al., 1990]. These observations widen the scope of TSS to encompass a group of diseases with similar pathophysiology and clinical manifestations which are caused by Ia-binding, superantigenic bacterial toxins.

Binding of TSST-1 to Ia Molecules

Evidence for the binding of staphylococcal exotoxins to Ia molecules was first provided by the demonstration that the mitogenic effect of these toxins is strictly dependent on the presence of Ia+ accessory cells in culture yet, unlike the case with nominal antigens, did not require intracellular processing [Fleischer and Schrezenmeier, 1988]. In the case of TSST-1, several lines of evidence were gathered indicating that this toxin binds to Ia molecules. First, TSST-1 was found to be mitogenic for peripheral blood mononuclear cells isolated from normal subjects but not from Ia-deficient patients, demonstrating a requirement for Ia+ accessory cells for TSST-1 mitogenicity similar to what was found for staphylococcal enterotoxins. TSST-1 was found to bind with high affinity (Kd_{50} ranging from 1.7 to 4.3×10^{-8}) to Ia+ but not Ia− human cell lines. The number of TSST-1 binding sites per cell correlated well with Ia surface density as measured by flow cytometry. Cultured human fibroblasts, which do not constitutively express Ia molecules, failed to bind TSST-1. Induction of Ia expression on fibroblasts by treatment with interferon-γ was associated with the appearance of TSST-1-binding sites. Binding of TSST-1 to Ia+ cells was inhibited by monoclonal antibodies recognizing monomorphic determinants on Ia molecules. Conclusive evidence for the binding of TSST-1 to Ia molecules was obtained by demonstrating that purified Ia molecules, but not MHC class I molecules, bind ^{125}I-labeled TSST-1 [Scholl and Geha, unpubl. observations].

TSST-1 can bind to Ia molecules of diverse isotypic and allotypic specificities. Our studies with L cells transfected with cDNA coding for α- and β-chains of different Ia isotypes (DP, DQ or DR) demonstrated that TSST-1 binds equally well and with high affinity to HLA-DR and DQ molecules [Scholl et al., 1990a]. No detectable binding could be demonstrated for HLA-DP. However, murine L cells transfected with HLA-DP supported the proliferation of T cells in response to TSST-1, suggesting that TSST-1 does indeed bind to HLA-DP molecules but with a much lower affinity than that observed for HLA-DR or DQ molecules. These results are similar to what has been found for other staphylococcal and streptococcal superantigens, which demonstrate an affinity pattern for Ia isotypes similar to what is noted for TSST-1 [Marrack and Kappler, 1990]. Studies on 14 different HLA-DR, 2 HLA-DQ and 2 HLA-DP alleles demonstrated no significant difference between the alleles of the same isotype in their affinity of binding to TSST-1. Also, all alleles tested were similar in their capacity to support the proliferation of human T lymphocytes in response to TSS1-1. These results suggested that Ia polymorphism would not play an important role in determining the response of immune cells of different individuals to TSST-1 and consequently the susceptibility of these individuals to TSST-1-mediated diseases. However, a different conclusion was reached in a study by

Herman et al. [1990]. These authors used IL-2 production by a TSST-1-responsive murine Vβ3 clone as a read-out system to study the influence of isotypic and allotypic specificities of Ia molecules on the activation of T lymphocytes by TSST-1. Strong differences emerged in the ability of different HLA-DR allotypes to present TSST-1 to this clone. HLA-DR1 provided the best response, while HLA-DR6, DR7 and DRw53 failed to present TSST-1. This directly contrasts with the results obtained by Scholl and co-workers, where for example DR1 and DR7 were equally efficacious in presenting TSST-1 to human T lymphocytes. The discrepancy between the two sets of results is likely to reside in the better capacity of human TCR molecules, as compared to murine TCR molecules, to recognize TSST-1 in complex with different alleles of human Ia molecules.

In addition to binding to human Ia molecules, TSST-1 has also been demonstrated to bind to murine Ia molecules [Scholl et al., 1990b; Uchiyama et al., 1989]. However, unlike the case with human Ia molecules, the binding of TSST-1 to murine Ia molecules is governed by both isotypic *and* allotypic specificities. Of the 2 murine Ia isotypes, TSST-1 binds well to I-A but not to I-E molecules. Within I-A, allelic differences in binding to TSST-1 have been observed. Thus TSST-1 binds well to I-A^b and to I-A^d but only weakly to I-A^k. The characteristics of TSST-1 binding to murine Ia molecules differed from those observed for other staphylococcal exotoxins such as SEA or SEB. For example, I-E molecules can support the proliferation of human T cells in response to SEA [Fleischer et al., 1989]. These results suggest that there exist important differences between the various superantigens in their ability to bind murine Ia molecules.

Superantigens bind to a site on Ia molecules outside of the antigen groove. Extensive mutations of Ia α1 residues that form one face of the antigen groove fail to affect the presentation of superantigens to T cells [Dellabona et al., 1990]. In contrast, the same mutations drastically affect the ability of Ia molecules to present nominal antigen to T cells. A similar conclusion was reached in another study on the binding of TSST-1 to Ia molecules, which failed to document an effect of TSST-1 on the binding of antigenic peptides to Ia molecules [Karp et al., 1990]. The same study took advantage of the disparity in the affinity of binding of TSST-1 to HLA-DR vs. HLA-DP molecules to map the binding site of TSST-1 on Ia molecules. A series of chimeric HLA DR/HLA DP molecules were generated and expressed in L cells, in which part of the sequence of either the α- or the β-chain was substituted by the corresponding sequence of the heterologous isotype. Thus, DR α-chains paired with DR/DP β-chains bound ^{125}I-TSST-1, whereas

DP α-chains paired with DP/DR β-chains did not. This suggested that it is the α-chain that determines the efficacy of binding of TSST-1 to Ia molecules. Furthermore, substituting the α2 domain of HLA-DR α-chain with that of HLA-DP did not diminish the binding of TSST-1, indicating that it is the α1 domain that is critical for binding of TSST-1 to Ia molecules. Another approach for mapping the binding site of TSST-1 to Ia molecules took advantage of the binding of TSST-1 to HLA-DR molecules but not to the highly homologous murine I-E molecules. L cell transfectants expressing hybrid DRα: I-Eβ, but not I-Eα-DR1β molecules, could bind ^{125}I-labeled TSST-1 [Scholl et al., 1990b]. This suggested that it is the α-chain specificity that is critical for the binding of TSST-1 to Ia molecules, in agreement with the results obtained using HLA-DR/DP hybrid molecules.

Different superantigens may bind to different sites on Ia molecules [Scholl et al., 1989b]. This has been most clearly demonstrated for TSST-1 and SEB. Neither toxin is capable of displacing the binding of the other toxin to B cell lines or to L cells transfected with Ia molecules. The capacity of murine I-E molecules to support SEA but not TSST-1-induced mitogenesis also argues for strong differences between TSST-1 and other toxins in their binding to Ia molecules.

The sites within TSST-1 responsible for binding to Ia molecules have not yet been mapped. However, there are clues as to their location. TSST-1 is a 22-kD protein composed of 194 amino acids [Blomster-Hautamaa et al., 1986]. Previous studies using cyanogen bromide generated fragments revealed that monoclonal antibodies that inhibited toxin-induced T cell proliferation reacted with a 14- to 15-kD internal fragment [Blomster-Hautamaa et al., 1986; Kokan-Moore and Bergdoll, 1989]. Further studies identified one epitope, a 10 amino acid stretch spanning residues 34 to 43, that is recognized by a neutralizing monoclonal antibody [Murphy et al., 1988]. Additional information was provided by studies on toxin fragments generated using the protease papain. Biological activity, as assayed by mitogenicity and by reactivity to neutralizing anti-TSST-1 monoclonal antibodies, was maintained in a 12-kD peptide spanning residues 88 to 194 of the toxin [Edwin et al., 1988]. It is reasonable to assume that this stretch contains both the Ia binding site as well as the site interacting with Vβ residues on the T cell receptor and that additional site(s) at the N-terminal region of the molecule may regulate the conformation of the molecule and provide sites for some neutralizing antibodies. Studies with overlapping peptides spanning the entire TSST-1 sequence at 20 amino acid stretches at a time have demonstrated that no one peptide can either inhibit or mimic the binding of TSST-

1 to Ia molecules [Ramesh and Geha, unpubl. observations]. This suggested that the Ia-binding site may be formed by a complex three-dimensional epitope not reproduced by any of the studied peptides.

Functional Consequences of Toxin/Ia Interaction

The binding of TSST-1 to Ia molecules initiates the activation of diverse immune cells including T and B lymphocytes, monocytes and natural killer cells. Two major mechanisms account for the activation of immune cells by TSST-1. The first mechanism accounts for the activation of T lymphocytes by TSST-1 and involves the engagement of the Vβ component of the T-cell receptor by toxin/Ia complexes resulting in T-cell activation and proliferation. The second mechanism involves the transduction via Ia molecules of signals that result in the activation of Ia+ immune cells including B lymphocytes, monocytes, activated natural killer cells and activated T lymphocytes.

Activation of T Lymphocytes. The mitogenic effect of TSST-1 is strictly dependent on the presence of Ia+ accessory cells. Radioligand binding assays fail to demonstrate binding of TSST-1 to Ia− T lymphocytes. Also, unlike the case of mitogenic anti T-cell receptor antibodies, TSST-1 is not mitogenic when cross-linked on plastic and presented to T lymphocytes in the presence of IL-2. Interestingly, purified Ia molecules cross-linked on plastic can support the proliferation of T lymphocytes in response to TSST-1 and IL-2 [Scholl and Geha, in preparation]. These results indicate that a specific, high-affinity interaction between TSST-1 and T-cell receptor molecules first requires the binding of TSST-1 to Ia molecules. In the presence of Ia+ accessory cells, TSST-1 induces the proliferation of T cells and the production of large quantities of lymphokines. However, not all T cells in culture respond to TSST-1. The uneven proliferative responses of different T cells to TSST-1 can be readily demonstrated using T-cell clones, some of which respond vigorously to TSST-1 while others fail to do so. The clonal variability between T cells in their mitogenic response to TSST-1 as well as other superantigens has been determined to be governed by the Vβ specificity of the responding T lymphocytes. In the case of TSST-1, only those lymphocytes expressing Vβ2 proliferate in response to TSST-1. This has been demonstrated to be true both in vitro, where treatment of peripheral blood mononuclear cells with TSST-1 results in the selective expansion of Vβ2 expressing T cells and in vivo, where patients with TSS demonstrate massive expansion of their Vβ2+ T lymphocyte population [Choi et al., 1990]. TSST-1 can also activate murine T cells in a Vβ-restricted manner [Marrack and

Kappler, 1990]. Murine T lymphocytes activated by TSST-1 include those bearing Vβ3, Vβ15 and Vβ17 products.

The molecular interactions governing the recognition of TSST-1/Ia complexes by Vβ2-compatible T-cell receptor molecules have not been specifically addressed. They may, however, be similar to what has been described for another staphylococcal enterotoxin, staphylococcal entero-toxin C2 (SEC2) [Choi et al., 1990]. SEC2/Ia complexes were recognized by chimeric murine T-cell receptor molecules bearing human Vβ13.2, but not Vβ13.1, product. This confirmed the permissive role of the T-cell receptor α-chain in the interaction of the β-chain with toxin/Ia complexes. Further studies identified the Vβ specificity of this toxin to be governed by a single stretch of eight amino acids facing away from peptide antigen/Ia binding site and found in Vβ13.2 but not in Vβ13.1. The interaction between TSST-1 and the T-cell receptor β-chain is contributed to by an additional interaction between the T-cell receptor and Ia molecules. This is supported by findings discussed in the previous section and demonstrating that TSST-1, bound by the same DR allele, e.g. DR7, is recognized by human but not by murine Vβ-compatible T lymphocytes. The site on TSST-1 recognized by the T-cell receptor is not known, but it clearly localizes to the distal two-thirds of the molecule as this portion of TSST-1 is fully mitogenic. The inability to demonstrate an interaction between the T-cell receptor and TSST-1 in the absence of Ia molecules suggests that the epitope on TSST-1 recognized by the T-cell receptor results from a conformational change in the toxin molecule that takes place upon its binding to Ia molecules.

Ia-bound TSST-1 activates both CD4+ and CD8+ T lymphocytes [Calvano et al., 1984], in marked distinction to Ia-bound nominal antigens which activate CD4+ cells only. The role of CD4 in the interaction of TSST-1/Ia complex with T-cell receptor molecules was examined using CD4– murine T-cell hybridomas [Sekaly et al., 1991]. Induction of CD4 expression by transfection with retroviral vectors containing CD4 cDNA did not affect the proliferative response of the majority of these hybridomas to TSST-1 or to other bacterial superantigens, indicating that CD4 is not required for the interaction between TSST-1/Ia complexes and T cell receptor molecules. This may reflect the high-affinity nature of TSST-1/Ia/TCR interaction, as CD4 expression is required in cases of low-affinity, but not high-affinity, interactions between nominal antigen Ia complexes and TCR molecules [Marrack et al., 1983].

Engagement of Vβ2 by TSST-1/Ia complexes results in the initiation of early activation events similar to those observed upon engagement of the T-cell receptor with nominal antigen/Ia complexes or with anti-T-cell receptor antibodies [Chatila et al., 1988; Norton et al., 1990]. These include the activation of T-cell receptor-coupled phospholipase C, which hydrolyzes membrane phosphoinositides to generate second messengers such as diacyl-glycerol, which activates protein kinase C, as well as inositol phosphates, which effect a rise in free intracellular Ca^{2+} concentration. These early activation events are thought to mediate the induction of lymphokine and lymphokine receptor gene expression as well as the progression of T cells through the cell cycle.

Ia molecules expressed on TSST-1-activated T lymphocytes can initiate further cycles of T cell activation and proliferation by presenting TSST-1 to T-cell receptor molecules. This has been demonstrated using Ia+ T cell clones bearing the appropriate Vβ product, which proliferate vigorously to TSST-1 in the absence of any added accessory cells [Spertini et al., 1991]. TSST-1 can be presented to T-cell receptor molecules by Ia molecules found on the same lymphocyte (cis presentation) or by Ia molecules present on other T lymphocytes (trans presentation). The induction by TSST-1 of sustained Ca^{2+} mobilization in these lymphocytes within seconds after its addition despite constant stirring suggests that cis presentation may be the dominant mechanism involved.

Recently, we have elucidated another mechanism that can help perpetu-ate the activation of T lymphocytes by TSST-1. This involves the delivery of trophic signals via Ia molecules that are expressed on activated T lympho-cytes and resulting in the activation and proliferation of these lymphocytes in a Vβ *unrestricted* manner [Spertini et al., 1991]. This has been demonstrated using Ia+ T-cell clones bearing a mismatched Vβ product. Unlike the Vβ-matched T-cell clones, the mismatched clones fail to proliferate to TSST-1 in the absence of accessory cells. However, they proliferate vigorously to TSST-1 in the presence of either Ia+ or Ia− accessory cells, suggesting that the activation signal is delivered via Ia molecules on T lymphocytes and requires an Ia-independent accessory cell signal. Further evidence for this scheme has been obtained using T-cell clones derived from Ia− deficient patients and lacking any detectable surface expression of Ia molecules. Unlike Ia+ T-cell clones, Ia− clones proliferate to TSST-1 only in a Vβ restricted manner. This indicated that the proliferation of Ia− clones can only be achieved by cognate interaction between Ia/toxin molecules on accessory cells and Vβ2+ T-cell receptor molecules on T lymphocytes.

Activation of B Lymphocytes. B lymphocytes express Ia molecules throughout much of their development and differentiation and are thus targets of TSST-1 binding and action. Postmortem studies on lymph nodes of patients with fatal TSS reveal evidence of both follicular and intrafollicular hyperplasia, suggesting in vivo activation and proliferation of follicular B lymphocytes. In vitro studies reveal that by itself TSST-1 does not cause the proliferation of B lymphocytes or their differentiation into immunoglobulin (Ig)-producing cells. However, in the presence of T lymphocytes, TSST-1 induces intense proliferation of B lymphocytes and promotes their differentiation into Ig secreting lymphocytes [Mourad et al., 1989]. TSST-1-induced Ig production is critically dependent on the presence of an optimal load of T lymphocytes in culture with B lymphocytes, estimated at 1:20 ratio of T to B lymphocytes. Interestingly, this ratio corresponds to the ratio of T to B lymphocytes normally found in lymph node follicles [Heinen et al., 1988]. Higher T lymphocyte loads result in progressive decline in TSST-1-induced Ig production. This may help explain the observation that TSST-1 inhibits spontaneous as well as pokeweed mitogen-induced Ig production in unfractionated peripheral blood mononuclear cells. It is possible that inhibitory T/B cell interactions dominate at higher T cell loads. Alternatively, massive activation of large numbers of T lymphocytes by TSST-1 may result in the attainment of a critical concentration of cytokine(s) inhibitory for TSST-1 triggered Ig production.

The mechanism by which TSST-1 induces T lymphocyte-dependent B lymphocyte proliferation and Ig production resides in its capacity to mediate cognate interaction between B lymphocytes and T lymphocytes. This interaction mimics the interaction between antigen-presenting B lymphocytes and antigen-responding T lymphocytes in its requirement for the participation of Ia molecules, T cell receptor/CD3 complexes and adhesion molecules especially LFA-1 and its counterreceptors. TSST-1 does not require the participation of CD4 molecules in its mediation of cognate interaction between T and B lymphocytes. As discussed in the previous section, this may reflect the high-affinity nature of the interaction between the Ia/toxin complex and the TCR/CD3 complex, as CD4 functions to bolster low affinity but not high-affinity antigen-TCR interactions. Another distinction between antigen and TSST-1 triggered T/B cell interactions is that the latter is not restricted to a particular Ia allele and can proceed between Ia-mismatched T and B lymphocytes. The ability of TSST-1 to mediate Ia-unrestricted cognate interaction between large numbers of unprimed T and B cells provides a useful model for the study of antigen-driven, T lymphocyte-dependent B lymphocyte proliferation and Ig production.

Polyclonal activation of B lymphocytes by TSST-1 can result in the inhibition of humoral immune responses to nominal antigens, including

TSST-1 itself. This may explain the inability of many TSS patients to mount an antibody response to TSST-1. As discussed earlier in this review, the majority of patients are found lacking antibodies to TSST-1 on their first presentation with TSS, suggesting that they have not previously encountered TSST-1 as a nominal antigen. The consequent polyclonal T and B lymphocyte activation may prevent them from mounting such a humoral immune response, making them susceptible to recurrent TSS. Supporting evidence for such a scenario has been provided by studies on a murine model of enterotoxin toxicity, where mice exhibited severely defective immune responses to a nominal antigen administered simultaneously with SEB.

While TSST-1 does not on its own induce the proliferation or differentiation of purified B lymphocytes, it does deliver activation signals to B lymphocytes via Ia molecules. TSST-1 synergizes with B cell mitogens such as phorbol myristate acetate (PMA) or anti-surface Ig antibodies in inducing the proliferation of B lymphocytes [Fuleihan et al., 1991]. Further evidence for signal transduction via Ia molecules in B lymphocytes has been provided by the demonstration that TSST-1 as well as monoclonal antibodies that recognize an epitope closely related to the TSST-1-binding site on Ia molecules induce sustained, LFA-1-dependent adhesion [Mourad et al., 1990]. This adhesion was effected by the activation of LFA-1 adhesion function and not that of its counterreceptors. Induction of cell adhesion by TSST-1 was also observed in Ia+ T lymphocytes and monocytes as well as Ia+, but not Ia−, B lymphoblastoid cell lines. This suggested that adhesion induced via Ia reflects a function common to Ia molecules in diverse cell types (see below).

Activation of NK Cells. Resting NK cells do not express Ia molecules on their surface and are thus not a target for TSST-1 action. Nevertheless, these cells, which constitutively express IL-2 receptors, can be potentially activated by IL-2 produced by TSST-1-activated T lymphocytes. Another mechanism for NK cell activation can proceed upon the induction of Ia expression by another lymphokine produced by TSST-1-activated T lymphocytes, namely interferon-γ. TSST-1 can then bind to and directly induce the proliferation of such Ia+ NK cells. This has been demonstrated using Ia+, CD56+, CD3− NK clones [Spertini et al., 1991]. TSST-1-induced proliferation of NK clones proceeded in the absence of accessory cells, but was enhanced upon the addition of accessory cells to cultures. It was inhibited by anti-Ia monoclonal antibodies that block the binding of TSST-1 to Ia molecules. These results suggested that TSST-1-induced proliferation of NK clones was triggered by signaling via Ia molecules.

Activation of Human Monocytes. TSST-1 and related superantigens are potent activators of human monocytes. The effects these toxins exert on monocytes include induction of homotypic cell adhesion, similar to what is described above for B lymphocytes, and induction of cytokine gene expression, described below.

Induction of Monokine Synthesis. The similarity between TSST-1 mediated toxic shock and endotoxin-mediated shock led investigators soon after the discovery of TSST-1 to examine whether it can mimic endotoxin in inducing IL-1 synthesis. It was quickly appreciated that TSST-1 is a potent inducer of IL-1 synthesis, more potent on a molar basis than endotoxin [Ikejima et al., 1984; Parsonnet et al., 1985, 1986]. TSST-1 was also demonstrated to induce the synthesis of TNF in human monocytes [Fast et al., 1989; Jupin et al., 1988; Parsonnet and Gillis, 1988]. Induction of IL-1 and TNF synthesis by TSST-1 does not require the presence of T lymphocytes as it can be demonstrated in TSST-1-treated monocytic cell lines [Hirose et al., 1985]. The presence of T lymphocytes may, however, enhance the efficacy of TSST-1 in inducing monokine synthesis [Fischer et al., 1990]. We have recently demonstrated that induction of IL-1 and TNF synthesis by TSST-1 is effected by transcriptional activation of the respective monokine gene, resulting in a dramatic increase in the level of newly initiated monokine RNA molecules and in monokine mRNA steady state levels in the nuclei and cytosol of stimulated monocytic cells, respectively [Trede et al., 1991]. Monokine mRNA species are first detected within 30 min after TSST-1 treatment; their levels peak at 3 h poststimulation and decline thereafter. Transcriptional activation of monokine genes by TSST-1 does not require prior protein synthesis as it proceeds in monocytes pretreated with cycloheximide, an inhibitor of protein synthesis. Monoclonal antibodies that recognize epitopes closely related to the TSST-1-binding site on Ia molecules also induce monokine mRNA accumulation. This is in agreement with previous reports demonstrating that some anti-Ia monoclonal antibodies induce IL-1 synthesis [Palacios, 1985; Palkama et al., 1989] and indicate that TSST-1 transduces activation signals via Ia molecules that result in the induction of monokine gene transcription.

Mechanisms of Signal Transduction by TSST-1 via Ia Molecules

Signal Transduction Events at the Cell Surface
Insight into the nature of activation signals transduced by TSST-1 via Ia was first gained by studying the effect of protein kinase inhibitors on toxin-

mediated induction of cell adhesion and monokine gene transcription. It was demonstrated that adhesion induced by TSST-1 and other Ia ligands was reversed by the protein kinase C-specific inhibitor sphingosine but not by its metabolite ceramide [Mourad et al., 1990]. This suggested that the binding of TSST-1 to Ia molecules induced intracellular activation signals that resulted in protein kinase C activation and consequent upregulation of LFA-1 adhesion function. Similarly, it was demonstrated that treatment of monocytes with the protein kinase inhibitors sphingosine or calphostin C or the tyrosine kinase inhibitors genistein or herbimycin inhibited the induction of monokine gene expression by TSST-1 [Trede, Geha and Chatila, submitted]. In contrast, inhibitors of cAMP-dependent protein kinases such as HA1004 did not affect the accumulation of IL-1 and TNF mRNA induced by TSST-1. These results suggested that transcriptional activation of monokine genes by TSST-1 is mediated by the activation of protein kinase C and of protein tyrosine kinase(s), acting either in series or in parallel.

Direct evidence for the induction by TSST-1 of enhanced tyrosine phosphorylation in target Ia+ cells was provided by probing cytosolic extracts of toxin-stimulated cells with antiphosphotyrosine-specific antibodies (Western blotting). Treatment of peripheral blood monocytes and of THP-1 monocytic leukemia cells with TSST-1 induced rapid increase in the levels of phosphotyrosine residues present in the cytosolic proteins of treated cells. Phosphotyrosine levels peaked at 30–60 s poststimulation and remained elevated for at least another 15 min. Induction by TSST-1 of enhanced tyrosine phosphorylation in monocytic cells is mediated by the activation of src type tyrosine kinases. Treatment of peripheral blood monocytes with TSST-1 induced rapid (30–60 s), intense activation of the p56^{fgr} and p59^{hck}, two src type tyrosine kinases whose expression is largely restricted to myelomonocytic cells (fig. 1). The specificity of this activation was established by the failure of TSST-1 to concurrently activate another src kinase, p59^{fyn}. The capacity of TSST-1 to activate p56^{fgr} and p59^{hck} in human monocytes was shared by other superantigens such as SEA. Finally, signaling by exotoxins via Ia molecules in B lymphocytes also resulted in the activation of src-type kinases, indicating that src-kinase activation is a feature of signaling via Ia molecules in different cell types [Morio, Geha and Chatila, in preparation].

Phosphorylation of phospholipase Cγ isoenzyme on tyrosine residues, leading to enzyme activation, is an event that commonly follows signaling via receptors endowed with or coupled to tyrosine kinase activity such as receptors for growth factors [Ullrich and Schlessinger, 1990] and T and B

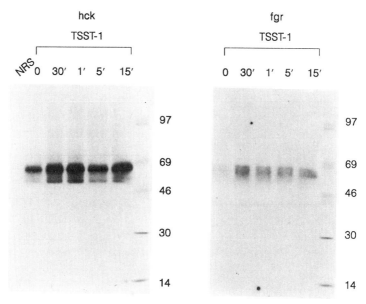

Fig. 1. TSST-1 induces rapid activation of p59^{hck} and p56^{fgr} in peripheral blood monocytes. Monocytes were stimulated with TSST-1 for the indicated time periods. Cells were then permeabilized with α-lysophosphatidylcholine and radiolabeled with γ^{32}p-ATP for 15 min on ice. Cells were then solubilized with 1% NP40 detergent lysis buffer and src related kinases were immunoprecipitated using either normal rabbit serum (NRS) or kinase-specific antisera.

lymphocyte receptors for antigen, respectively [Bolen, 1991]. As discussed above, signaling by TSST-1 via the T cell receptor is rapidly followed by the activation of phospholipase C. The generation by this enzyme of the phosphoinositides-derived second messengers diacylglycerol and inositol phosphates induces the translocation of protein kinase C (PKC) from cytosol to cellular particulate fractions and a rise in intracellular free Ca^{2+} concentrations, respectively. A similar pattern of events is observed upon signaling by TSST-1 and related exotoxins via Ia molecules in monocytic cells. It thus appears that Ia molecules act to mediate cellular activation events in a manner similar to the case of other members of the immunoglobulin supergene family such as lymphocyte receptors for antigen. For these receptors, including Ia molecules, activation of src-type tyrosine kinases and PKC plays a central role in transducing growth-promoting signals intracellularly (presented schematically in fig. 2). As is the case for antigen receptors, it seems likely that there exists a host of Ia-associated molecules that act as signal

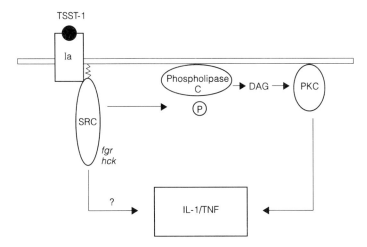

Fig. 2. Schematic representation of Ia-coupled signal transduction pathways activated by TSST-1 and involved in induction of monokine gene transcription.

transducers and that may interact with src kinases. The composition and organization of such an Ia-signal transducing complex remains to be established.

Activation of Transcriptional Factors

Transcriptional activation of cytokine genes is mediated by a set of DNA
binding factors that recognize distinct regulatory sequences found within the
promotor and enhancer regions of these genes [Crabtree, 1989]. Examination
of IL-1β [Clark et al., 1986], TNFα and TNFβ [Nedospasov et al., 1986]
genomic sequences reveals DNA sequences recognized by AP-1, AP-2 and
NF-κB DNA binding proteins. Activation of NF-κB has been particularly
implicated as a mechanism for induction of cytokine gene expression and of
HIV gene expression by various stimuli such as LPS [Shakhov et al., 1990].
Recent studies carried out in our laboratory have established TSST-1 and
related exotoxins as potent activators of NF-κB as well as other transcriptional factors such as AP-1 [Trede, Geha and Chatila, in preparation]. NF-κB
activation is induced within 5 min following TSST-1 addition, as monitored
by electrophoretic mobility gel shift assays, and peaks by 30–60 min thereafter. TSST-1 and related staphylococcal superantigens potently activate the
transcription of reporter genes coupled to an NF-κB enhancer sequence and
transfected into the human monocytic cell line THP-1. Activation of NF-κB
by superantigens is antagonized by inhibitors of protein kinase C and of

protein tyrosine kinases, similar to what was observed for the induction of monokine gene expression by these toxins. These findings point to a cascade of cytoplasmic and nuclear activation events that follows the engagement of MHC class II molecules by superantigens. This cascade entails the activation of transcriptional factors by toxin-stimulated intracellular protein kinases, leading to the induction of gene transcription.

Toxic Shock Syndrome and Endotoxin-Mediated Shock: Common and Divergent Pathogenic Mechanisms

As alluded to previously, the pathophysiology of TSS bears resemblance to LPS-mediated gram negative bacterial sepsis. In each of these conditions, the pathogenic bacterial product induces the release of monokines such as IL-1 and TNF implicated in causing tissue injury. Induction of monokine synthesis by TSST-1 and LPS may proceed via a common mechanism, with LPS activating intracellular signaling mechanisms normally coupled to Ia molecules and amenable to activation by extracellular ligands such as TSST-1. IL-1 [Okusawa et al., 1988] and TNF [Tracey et al., 1986] precipitate shock when infused into experimental animals, and the combination of both monokines exhibits synergy in inducing shock [Ikejima et al., 1989; Okusawa et al., 1988]. Conversely, anti-TNF antibodies protect experimental animals from endotoxin-mediated shock [Tracey et al., 1987]. It is thus likely that monokine release helps mediate the shock state observed in both conditions. However, unlike endotoxin-mediated sepsis, TSS is additionally associated with the activation of a significant portion of T lymphocytes as a result of superantigen-mediated activation by TSST-1 and related toxins. The consequent outpouring of lymphokines may synergize with monokines to aggravate the shock state. Activated lymphocytes with enhanced adhesive properties may interact with vascular endothelium to induce inflammation, end-organ damage and vascular permeability. The ability of different staphylococcal exotoxins to activate populations of T cells bearing distinct Vβ products would result in an even greater degree of T cell activation by staphylococcal isolates producing multiple toxins. Such isolates may be associated with a higher fatality rate, as has been observed for those isolates producing TSST-1 and SEC1 [Crass and Bergdoll, 1986]. T lymphocyte activation may also account for some of the peculiar manifestations of TSS such as generalized erythroderma, a condition that may reflect intense lymphocyte activation and infiltration around small blood vessels in the skin, and periportal lymphocytic infiltration in the liver. The powerful effects of TSST-1 on the immune system may account for the tenacity of the shock state that is observed in otherwise young and healthy individuals and which is mediated at times by very small pockets of staphylococcal infection.

Homology between TSST-1 and Heat Shock Proteins: Its Potential Role in Immunity to Toxin and in Toxin-Triggered Autoimmune Diseases

Like any other protein antigen, superantigens are subject to intracellular processing by antigen-presenting cells. Peptides derived from superantigens could be presented to T cells as conventional antigens, promoting the emergence of toxin-specific cellular immunity. Most importantly, cognate interaction between toxin-specific helper T lymphocytes and toxin-peptide-presenting B cells would allow for the emergence of protective humoral immunity. To examine the capacity of the immune system to respond to TSST-1 as a conventional protein antigen, a panel of adult donors were screened for the capacity of their peripheral blood lymphocytes to respond to a series of 10 overlapping peptides derived from the TSST-1 sequence. None of the peptides behaved as a superantigen as judged by the following criteria: they did not compete with TSST-1 for binding to MHC class II molecules, they did not induce the proliferation of naive T cells such as mature thymocytes or cord blood lymphocytes, cells which otherwise respond vigorously to TSST-1, and they failed to exhibit the biological effects of superantigens such as induction of monokine gene expression. However, most of these peptides did induce the proliferation of peripheral blood lymphocytes derived from at least some of the donors tested. Two peptides, peptide 1 (amino acid 18–37) and peptide 2 (amino acid 34–63) induced the proliferation of peripheral blood lymphocytes from the majority of subjects, and this proliferation was inhibited by monoclonal antibodies to HLA DR but not to HLA DP or DQ [Ramesh et al., 1992a, b]. Lymphocyte proliferation to both peptides was not restricted to particular HLA DR allotypes; both peptides were, however, recognized by T cells when presented by autologous HLA DR types. These results suggested that these 2 peptides are 'universally' presented by HLA DR alleles, similar to what has been described for tetanus toxoid and malarial circumsporozite peptides.

Search for homology between peptide 1 and peptide 2 sequences and those of other proteins revealed significant homology to heat shock proteins sequences: peptide 1 was homologous to an hsp 65 peptide spanning residues

I HSP 65

			*	*	*	*	*							*		*	*	*
TSST-1 (18–37)	T	F	T	N	S	E	V	L	D	N	S	L	G	S	M	R	I
M. leprae hsp 65 (180–196)	T	F	G	L	–	Q	L	E	L	T	E	–	G	–	M	R	F
Mammalian hsp 65 (205–221)	T	L	N	D	–	E	L	E	I	I	E	–	G	–	M	R	F
E. coli Gro EL (181–197)	G	L	Q	D	–	E	L	D	V	V	E	–	G	–	M	Q	F

II HSP 18

					*		*	*		*		*		*		
TSST-1 (34–63)	R	I	K	–	–	N	T	D	G	S	I	S	L	I	I	F	P
M. leprae hsp 18 (101–117)	R	I	L	A	S	Y	Q	<u>E	G	V	L	K	L	S</u>	I	P	V
SEB (205–221)	R	V	–	–	–	F	E	<u>D	G	K	N	L	L	S</u>	F	D	V

Fig. 3. Homology between TSST-1 and hsp proteins. Sequences are aligned to give best homology. *Residues that are either conserved or conservatively substituted. Underlined hsp 18 sequence represents an 8 amino acid sequence conserved among low-molecular-weight hsp.

180–196, while peptide 2 was homologous to hsp 18 peptide spanning residues 101–115 (fig. 3). Both hsp peptides have been implicated in the immune response to hsp. In the case of hsp 65 peptide 180–196, this peptide has been demonstrated to be the epitope recognized by hsp-reactive γδ T lymphocytes [Born et al., 1990]. In the case of hsp 18 peptide 101–115, this peptide contains an 8 amino acid core region that has been shown to be highly conserved among low-molecular-weight heat shock proteins derived from diverse sources [Young, 1988] (fig. 3). Interestingly, this core region is particularly conserved in TSST-1, and is also found in SEB (fig. 3) as well as in SEA, SED, and SEE (data not shown).

Evidence suggests that in the case of both peptides 1 and 2 of TSST-1, the sequence homology to hsp is functionally significant. For example, it could be demonstrated that the eight amino acid sequence of peptide 2 that was homologous to the hsp 18 core region of *Mycobacterium leprae* is critical for induction of lymphocyte proliferation. Furthermore, the mean proliferative response of lymphocytes derived from PPD-positive subjects, who are presumably sensitized to mycobacterial hsp 18, is significantly higher than that of lymphocytes derived from PPD-negative individuals. In the case of peptide 1, it could be demonstrated that this peptide induced the proliferation of T cells specific for tuberculin PPD. A high percentage of these cells are known to react to hsp 65, which is present in tuberculin PPD preparations. In contrast, peptide 1 fails to induce the proliferation of T cells specific for tetanus toxoid.

It is tempting to speculate that T cells primed to cross-reactive peptides from hsp may provide helper function in mounting an antibody response to TSST-1 and to other enterotoxins that contain hsp-like sequences. Activation by TSST-1 peptides of T cells primed to cross-reactive peptides from hsp may also provide a mechanism for the induction of autoimmune diseases in the wake of infections by enterotoxin-producing organisms. The immune response to hsp has been implicated in the pathophysiology of autoimmune diseases. This has been best illustrated for adjuvant-induced arthritis in rats, a model arthritic disease which is induced by immunization to mycobacteria. Disease onset correlates with development of reactivity to mycobacterial hsp 65, particularly to hsp 65 peptide 180–186 (which is homologous with TSST-1 peptide 18–37), and preimmunization of rats with hsp 65 prevents the onset of arthritis. It should thus be considered that priming of T cells with TSST-1/enterotoxin peptides in a genetically susceptible host could result in the initiation of an immune response against self antigens such as hsp leading to autoimmune manifestation.

Activation of HIV Expression

Patients afflicted with AIDS are frequently infected with superantigen-producing microorganisms. In addition, recent evidence suggests that HIV itself may encode superantigen [Imberti et al., 1991], as has been described for murine leukemia virus, the causative agent of murine AIDS [Hugin et al., 1991]. Bacterial superantigens may act via the T-cell receptor to induce cell anergy and programmable cell death (apoptosis). TSST-1 and other superantigens may additionally transmit activation signals via Ia molecules that activate HIV gene expression and viral protein secretion in Ia+ target cells. This latter possibility was further pursued by examining the capacity of staphylococcal exotoxins to activate HIV expression in human monocytic cells [Fuleihan et al., 1992]. These model Ia+ cells were used as they are a prime target of HIV infection, represent a major reservoir for the virus and play an important role in the propagation and dissemination of the virus. TSST-1 and related staphylococcal superantigens were found to potently activate the transcription of chloramphenicol acetyl transferase gene coupled to HIV long-terminal repeat and transfected into the human monocytic cell line THP-1 (fig. 4). These toxins also increased viral protein secretion from chronically infected monocytic cell line U1. Superantigen-induced activation of HIV expression was mediated in part by toxin-stimulated TNF-α secretion as it was partially inhibited by neutralizing antibodies. These results indicate that superantigens produced by microorganisms as well as putative super-

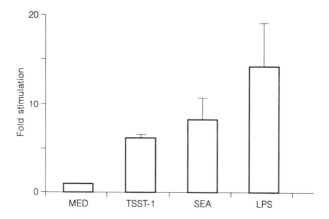

Fig. 4. Induction of HIV-1-CAT by staphylococcal exotoxins. TSST-1 and SEA activate human immunodeficiency virus (HIV-1) long-terminal repeat-driven transcription of chloramphenicol acetyl transferase gene (CAT), used as a reporter gene. HIV-1 CAT constructs were transfected into THP-1 human monocytic cells and stimulated with the respective reagent at 5 µg/ml overnight. LPS was used as a positive control. Results represent mean of three experiments ± SD.

antigens encoded by HIV may activate latent HIV infection in monocytic cells as well as in B cells and Ia+ T cells in patients infected with HIV, leading to disease progression.

Future Directions

The discovery that TSST-1 serves as an Ia-binding superantigen has greatly clarified the mechanisms by which this and other related staphylococcal exotoxins induce disease. This will hopefully allow for the development of strategies for countering the effects of these toxins on the immune system, including the design of agents that interfere with the binding of toxin to Ia molecules or that interrupt the massive cytokine release induced by these toxins and mediating much of their pathologic effects. The ready availability of TSST-1 as a high-affinity, agonistic Ia ligand provides a tool that can help unravel transmembrane signaling processes via Ia molecules. This will greatly expand our information on this important receptor system and the role it plays in the course of a normal immune response and in disease states.

References

Bergdoll MS, Crass BA, Reiser RF, Robbins RN, Davis JP: A new staphylococcal enterotoxin, enterotoxin F, associated with toxic shock syndrome Staphylococcal aureus isolates. Lancet 1981;i:1017–1021.

Best GK, Scott DF, Kling JM, Thompson MR, Adinolfi LE, Bonventre PF: Protection of rabbits in an infection model of toxic shock syndrome by a TSS toxin 1-specific monoclonal antibody. Infect Immun 1988;56:998–999.

Blomster-Hautamaa DA, Kreiswirth BN, Kornblum JS, Novick RP, Schlievert PM: The nucleotide and partial amino acid sequence of toxic shock syndrome toxin 1. J Biol Chem 1986;261:15783–15786.

Blomster-Hautamaa DA, Novick RP, Schlievert PM: Localization of biological function of toxic shock syndrome toxin-1 by use of monoclonal antibodies and cyanogen bromide generated toxin fragment. J Immunol 1986;137:3572–3576.

Bolen JB: Signal transduction by the src family of tyrosine protein kinases in hemopoietic cells. Cell Growth Differ 1991;2:409–414.

Bonventre PF, Thompson MR, Adinolfi LE, Gillis ZA, Parsonnet J: Neutralization of toxic shock syndrome toxin-1 by monoclonal antibodies in vitro and in vivo. Infect Immun 1988;56:135–141.

Bonventre PF, Wekbach L, Harth G, Haidaris C: Distribution and expression of toxic shock syndrome toxin 1 gene among *Staphylococcus aureus* isolates of toxic shock syndrome and non-toxic shock syndrome origin. Rev Infect Dis 1989;11(suppl 1): S90–S95.

Born W, Hall L, Dallas A, Boymel J, Shinnick T, Young D, Brennan P, O'Brien R: Recognition of a peptide antigen by heat shock-reactive γδ T lymphocytes. Science 1990;249:67–69.

Calvano SE, Quimby FW, Antonacci AC, Reiser RF, Bergdoll MS, Dineen P: Analysis of the mitogenic effects of toxic shock toxin on human peripheral blood mononuclear cells in vitro. Clin Immunol Immunopathol 1984;33:99–110.

Chatenoud L, Ferran C, Legendre C, Thouard I, Merite S, Reuter A, Gevaert Y, Kreis H, Franchimont P, Bach JF: In vivo cell activation following OKT3 administration. Systemic cytokine release and modulation by corticosteroids. Transplantation 1990; 49:697–702.

Chatila T, Wood N, Parsonnet J, Geha RS: Toxic shock syndrome toxin-1 induces inositol phospholipid turnover, protein kinase C translocation, and calcium mobilization in human T cells. J Immunol 1988;140:1250–1255.

Chesney PJ: Clinical aspects and spectrum of illness of toxic shock syndrome: Overview. Rev Infect Dis 1989;11(suppl 1):S1–S7.

Choi Y, Herman A, DiGiusto D, Wade T, Marrack P, Kappler J: Residues of the variable region of the T cell receptor β chain that interact with *S. aureus* toxin superantigens. Nature 1990;346:471–473.

Choi Y, Kotzin B, Herron L, Callahan J, Marrack P, Kappler J: Interaction of *Staphylococcus aureus* toxin superantigens with human T cells. Proc Natl Acad Sci USA 1989; 86:8941–8945.

Choi Y, Lafferty JA, Clements JR, Todd JK, Gelfand EW, Kappler J, Marrack P, Kotzin BL: Selective expansion of T cells expressing Vβ2 in toxic shock syndrome. J Exp Med 1990;172:981–984.

Clark BD, Collins KL, Gandy MS, Webb AC, Auron PE: Genomic sequence for human prointerleukin 1 beta: Possible evolution from a reverse transcribed prointerleukin 1 alpha gene. Nucl Acids Res 1986;14:7897–7914.

Cone LA, Woodward DR, Schlievert PM, Tomory GS: Clinical and bacteriological observations of a toxic shock-like syndrome due to Streptococcus pyogenes. N Engl J Med 1987;317:146–149.

Crabtree G: Contingent genetic regulatory events in T lymphocytes. Science 1989;243: 355–361.

Crass BA, Bergdoll MS: Toxin involvement in toxic shock syndrome. J Infect Dis 1986; 153:918–926.

Davis JP, Chesney PJ, Wand PJ, LaVenture M: Toxic-shock syndrome: Epidemiological features, recurrence, risk factors and prevention. N Engl J Med 1980;303:1429–1435.

de Azavedo JCS, Foster TJ, Hartigan PJ, Arbuthnott JP, O'Reilly M, Kreiswirth BN, Novick RP: Expression of the cloned toxic shock syndrome toxin 1 gene (tst) in vivo with a rabbit uterine model. Infect Immun 1985;50:304–309.

Dellabona P, Peccoud J, Kappler J, Marrack P, Benoist C, Mathis D: Superantigens interact with MHC class II molecules outside of the antigen groove. Cell 1990;62: 1115–1121.

Edwin C, Parsonnet J, Kass EH: Structure-reactivity relationship of toxic shock syndrome toxin 1: Derivation and characterization of immunologically and biologically active fragments. J Infect Dis 1988;158:1287–1295.

Fast DJ, Schlievert PM, Nelson RD: Toxic shock syndrome-associated staphylococcal and streptococcal pyrogenic toxins are potent inducers of tumor necrosis factor production. Infect Immun 1989;57:291–294.

Fischer H, Dohlsten M, Anderson U, Hedlund G, Ericson P, Hanson J, Sjøgren HO: Production of TNF-α and TNF-β by staphylococcal enterotoxin A activated human T cells. J Immunol 1990;144:4663–4669.

Fischer H, Dohlsten M, Lindvall M, Sjøgren HO, Carlsson R: Binding of staphylococcal enterotoxin A to HLA-DR on B cell lines. J Immunol 1989;142:3151–3157.

Fleischer B, Schrezenmeier H: T cell stimulation by staphylococcal enterotoxins. J Exp Med 1988;167:1697–1707.

Fleischer B, Schrezenmeier H, Conradt P: T lymphocyte activation by staphylococcal enterotoxins: role of class II molecules and T cell surface structures. Cell Immunol 1989;120:92–101.

Fraser JD: High-affinity binding of staphylococcal enterotoxins A and B to HLA-DR. Nature 1989;339:221–223.

Fuleihan R, Mourad W, Geha RS, Chatila T: Engagement of MHC-class II molecules by the staphylococcal exotoxin TSST-1 delivers a progression signal to mitogen activated B cells. J Immunol 1991;146:1661–1666.

Fuleihan R, Trede N, Chatila T, Geha RS: Superantigens activate HIV-1 gene expression in monocytes. 1992;submitted.

Galelli A, Anderson S, Charlot B, Alouf JE: Induction of murine hemopoietic growth factors by toxic shock syndrome toxin-1. J Immunol 1989;142:2855–2863.

Heinen E, Cormann N, Kinet-Denoel C: The lymph follicle: A hard nut to crack. Immunol Today 1988;9:240–243.

Herman A, Croteau G, Sekaly RP, Kappler J, Marrack P: HLA-DR alleles differ in their ability to present staphylococcal enterotoxins to T cells. J Exp Med 1990;172:709–717.

Hirose A, Ikejima T, Gill M: Established macrophagelike cell lines synthesize interleukin-1 in response to toxic shock syndrome toxin. Infect Immun 1985;50:765–770.

Hugin AW, Vacchio MS, Morse HC III: A virus-encoded 'Superantigen' in a retrovirus-induced immunodeficiency syndrome of mice. Science 1991;252:424–427.

Ikejima T, Dinarello CA, Gill DM, Wolff SM: Induction of human interleukin-1 by a product of *Staphylococcus aureus* associated with toxic shock syndrome. J Clin Invest 1984;73:1312–1320.

Ikejima T, Okusawa S, Van Der Meer WM, Dinarello CA: Toxic shock syndrome is mediated by interleukin 1 and tumor necrosis factor. Rev Infect Dis 1989;11(suppl 1):S316–S317.

Imanishi K, Igarashi H, Uchiyama T: Activation of murine T cells by streptococcal pyrogenic exotoxin type A: Requirement for MHC class II molecules on accessory cells and identification of Vβ elements in T cell receptor of toxin-reactive T cells. J Immunol 1990;145:3170–3176.

Imberti L, Sottini A, Bettinardi A, Puoti M, Primi D: Selective depletion in HIV infection of T cells that bear specific T cell receptor Vβ sequences. Science 1991;254:860–862.

Jupin C, Anderson S, Damais C, Alouf J: Toxic shock syndrome toxin 1 as an inducer of human tumor necrosis factor and γ interferon. J Exp Med 1988;167:752–761.

Kappler J, Kotzin B, Herron L, Gelfand EW, Bigler RD, Boylston A, Carrel S, Posnett DN, Choi Y, Marrack P: Vβ-specific stimulation of human T cells by staphylococcal toxins. Science 1989;244:811–813.

Karp DR, Teletski CL, Scholl P, Geha R, Long EO: The α1 domain of the HLA-DR molecule is essential for high-affinity binding of the toxic shock syndrome toxin-1. Nature 1990;346:474–476.

Kokan-Moore NP, Bergdoll MS: Determination of the biologically active region in toxic shock syndrome toxin 1. Rev Infect Dis 1989;11(suppl 1):S125–S126.

Marrack P, Blackman M, Kushner E, Kappler J: The toxicity of staphylococcal enterotoxin B in mice is mediated by T cells. J Exp Med 1990;171:455–464.

Marrack P, Enders R, Shimonkevitz R, Zlotnik A, Dialynas D, Fitch F, Kappler J: The major histocompatibility complex-restricted antigen receptor on T cells. II. Role of the L3T4 product. J Exp Med 1983;158:1077–1091.

Marrack P, Kappler J: The staphylococcal enterotoxins and their relatives. Science 1990; 248:705–711.

Micusan VV, Mercier G, Bahtti AR, Reiser RF, Bergdoll MS: Production of human and murine interleukin-2 by toxic shock syndrome toxin-1. Immunology 1986;58:203–207.

Mollick JA, Cook RG, Rich RR: Class II MHC molecules are specific receptors for staphylococcus enterotoxin A. Science 1989;244:817–820.

Mourad W, Geha RS, Chatila T: Engagement of major histocompatibility complex class II molecules induces sustained, LFA-1 dependent cell adhesion. J Exp Med 1990;172: 1513–1516.

Mourad W, Scholl P, Diez A, Geha R, Chatila T: The staphylococcal toxin TSST-1 triggers B cell proliferation and differentiation via MHC unrestricted cognate T/B cell interaction. J Exp Med 1989;170:2011–2022.

Murphy BG, Kreiswirth BN, Novick RP, Schlievert PM: Localization of a biologically important epitope on toxic shock syndrome toxin 1. J Infect Dis 1988;158:549–555.

Nedospasov SA, Shakov AN, Turetskaya RL, Mett VA, Azizov MM, Georgiev GP, Korobko VG, Dobrynin VN, Filippov SA, Bystrov NS, Boldyreva EF, Chuvpilo SA,

Chumakov AM, Ovchinnikov YA: Tandem arrangement of genes coding for tumor necrosis factor (TNF-α) and lymphotoxin (TNF-β) in the human genome. Cold Spring Harbor Symp Quant Biol 1986;511:611.

Norton SD, Schlievert PM, Novick RP, Jenkins MK: Molecular requirements for T cell activation by the staphylococcal toxic shock syndrome toxin-1. J Immunol 1990;144: 2089–2095.

Okusawa S, Gelfand JA, Ikejima T, Connolly RJ, Dinarello CA: Interleukin 1 induces a shock-like state in rabbits. Synergism with tumor necrosis factor and the effect of cyclooxygenase inhibition. J Clin Invest 1988;81:1162–1172.

Palacios R: Monoclonal antibodies against human Ia antigens stimulate monocytes to secrete interleukin 1. Proc Natl Acad Sci USA 1985;82:6652–6656.

Palkama T, Sihvola M, Hurme M: Induction of interleukin 1α (IL-1α) and IL-1β mRNA expression and cellular IL-1 production by anti-HLA-DR antibodies in human monocytes. Scand J Immunol 1989;29:609–615.

Paris AL, Herwaldt LA, Blum D, Schmid GP, Shands KN, Broome CV: Pathologic findings in twelve fatal cases of toxic shock syndrome. Ann Intern Med 1982;96:852–857.

Parsonnet J: Mediators in the pathogenesis of toxic shock syndrome: An overview. Rev Infect Dis 1989;11(suppl. 1):S263–S269.

Parsonnet J, Gillis ZA: Production of tumor necrosis factor by human monocytes in response to toxic shock syndrome toxin-1. J Infect Dis 1988;158:1026–1033.

Parsonnet J, Gillis ZA, Pier GB: Induction of interleukin-1 by strains of Staphylococcus aureus from patients with nonmenstrual toxic shock syndrome. J Infect Dis 1986; 154:55–63.

Parsonnet J, Gillis ZA, Richter AG, Pier GB: A rabbit model of toxic shock syndrome that uses a constant, subcutaneous infusion of toxic shock syndrome toxin 1. Infect Immun 1987;55:1070–1076.

Parsonnet J, Hickman RK, Eardley DP, Pier GB: Induction of human interleukin-1 by toxic shock syndrome toxin-1. J Infect Dis 1985;151:514–522.

Ramesh N, Ahern D, Geha RS: A toxic shock syndrome toxin-1 derived peptide that shows homology to mycobacterial heat shock protein 65 is cross recognized by Mycobacterium tuberculosis PPD sensitized T cells. 1992a;submitted.

Ramesh N, Spertini F, Scholl P, Geha RS: A toxic shock syndrome toxin-1 peptide that shows homology to mycobacterial heat shock protein 18 is presented as a conventional antigen to T cells by multiple HLA-DR alleles. J Immunol 1992b;148:1025–1030

Rasheed JK, Arko RJ, Feely JC, Chandler FW, Thornsberry C, Gibson RJ, Cohen ML, Jeffries CD, Broome CV: Acquired ability of Staphylococcus aureus to produce toxic shock associated protein and resulting illness in a rabbit model. Infect Immun 1985; 47:598–604.

Rosenberg SA, Lotze MT, Mule JJ: New approaches to the immunotherapy of cancer using interleukin-2. Ann Intern Med 1988;108:853–864.

Schlievert PM: Staphylococcal enterotoxin B and toxic-shock syndrome toxin-1 are significantly associated with non-menstrual TSS (letter). Lancet 1986;i:1149–1150.

Schlievert PM, Shands KN, Dan BB, Schmid GP, Nishimura RD: Identification and characterization of an exotoxin from Staphylococcus aureus associated with toxic shock syndrome. J Infect Dis 1981;143:509–516.

Scholl P, Diez A, Mourad W, Parsonnet J, Geha RS, Chatila T: Toxic shock syndrome toxin-1 binds to class II major histocompatibility molecules. Proc Natl Acad Sci USA 1989a;86:4210–4214.

Scholl PR, Diez A, Geha RS: Staphylococcal enterotoxin B and toxic shock syndrome toxin-1 bind to distinct sites on HLA-DR and HLA-DQ molecules. J Immunol 1989b;143:2583.

Scholl PR, Diez A, Karr R, Sekaly RP, Trowsdale J, Geha RS: Effects of isotypes and allelic polymorphism on the binding of staphylococcal exotoxins to MHC class II molecules. J Immunol 1990a;144:226–230.

Scholl PR, Diez A, Sekaly RP, Glimcher L, Geha RS: Binding of toxic shock syndrome toxin-1 to murine MHC class II molecules. Eur J Immunol 1990b;20:1911–1916.

Sekaly RP, Croteau G, Bowman M, Scholl P, Burakoff S, Geha RS: The CD4 molecule is not always required for the T cell response to bacterial enterotoxins. J Exp Med 1991.

Shakhov AN, Collart MA, Vassalli P, Nedospasov SA, Jongeneel CV: κB-type enhancers are involved in lipopolysaccharide-mediated transcriptional activation of the tumor necrosis factor α gene in primary macrophages. J Exp Med 1990;171:35–47.

Spertini F, Spits H, Geha RS: Staphylococcal toxins polyclonally activate major histocompatibility complex class II+ human T cell clones. Proc Natl Acad Sci USA 1991; 88:7533–7537.

Todd J, Fishaut M, Kapral F, Welch T: Toxic-shock syndrome associated with phage-group 1 staphylococci. Lancet 1978;ii:1116–1118.

Tracey JT, Fong Y, Hesse DG, Manogue KR, Lee AT, Kuo GC, Lowry SF, Cerami A: Anti-cachectin/TNF monoclonal antibodies prevent septic shock during lethal bacteraemia. Nature 1987;330:662–664.

Tracey KJ, Beutler B, Lowry SF, Merryweather J, Wolpe S, Milsark IW, Hairi IJ, Fahey TJ, Zentella A, Albert JD, Cerami A: Shock and tissue injury induced by recombinant human cachectin. Science 1986;234:470–473.

Trede N, Geha RS, Chatila T: Transcriptional activation of monokine genes by MHC class II ligands. J Immunol 1991;146:2310–2315.

Uchiyama T, Imanishi K, Saito S, Araake M, Yan X J, Fujikawa H, Igarashi H, Kato H, Obata F, Kashiwagi N, Inoko H: Activation of human T cells by toxic shock syndrome toxin-1: the toxin-binding structures expressed on human lymphoid cells acting as accessory cells are HLA class II molecules. Eur J Immunol 1989;19:1803–1809.

Uchiyama T, Kamagata Y, Wakai M, Yoshioka M, Fujikawa H, Igarashi H: Study of the biological activities of toxic shock syndrome toxin-1. I. Proliferative response and interleukin-2 production by T cells stimulated with the toxin. Microbiol Immunol 1986;30:469–483.

Uchiyama T, Tadakuma T, Imanishi K, Araake M, Saito S, Yan X-J, Fujikawa H, Igarashi H, Yamaura N: Activation of murine T cells by toxic shock syndrome toxin-1. The toxin-binding structure expressed on murine accessory cells are MHC class II molecules. J Immunol 1989;143:3175–3182.

Ullrich A, Schlessinger J: Signal transduction by receptors with tyrosine kinase activity. Cell 1990;61:203–212.

Young DB: Stress-induced proteins and the immune response to leprosy. Microb Sci 1988;5:143.

Talal Chatila, MD, Division of Immunology, The Children's Hospital,
Boston, MA 02115 (USA)

Fleischer B (ed): Biological Significance of Superantigens.
Chem Immunol. Basel, Karger, 1992, vol 55, pp 172–184

T-Cell-Dependent Shock Induced by a Bacterial Superantigen

Thomas Miethke, Helmut Gaus, Claudia Wahl, Klaus Heeg,
Hermann Wagner

Institute of Medical Microbiology and Hygiene, Technical University of
Munich, FRG

Introduction

Several bacterial exotoxins have been recently recognized as members of
a new class of antigens termed superantigens. These bacterial superantigens
as well as gene products of endogenous retroviruses of the mouse share the
ability to bind to nonpolymorphic domains of class II major histocompatibility (MHC) molecules regardless of their haplotype [1, 2] and to activate
T cells, expressing a certain Vβ gene segment of the T cell receptor (TCR) [3].
Depending on the Vβ distribution in a T cell population, superantigens thus
stimulate a large part of the peripheral T cell pool. The α chain of the TCR
seems not to participate in the recognition of superantigens [4]. The superantigen group consists of several toxins produced by *Staphylococcus aureus*
such as the enterotoxins (SEA, SEB, SEC1–3, SED, SEE), the toxic shock
syndrome toxin 1 (TSST-1) and the exfoliative toxins A and B [5, 6]. In
addition, the erythrogenic toxins A, B and C produced by *Streptococcus
pyogenes* and a soluble mitogen from *Mycoplasma arthritidis* were recognized as superantigens [5, 7]. At least one of these toxins, i.e. TSST-1, is
associated with shock or shock-like symptoms in humans [8, 9]. The toxic
shock syndrome is characterized by high fever, hypotension, generalized
erythroderma and dysfunction of several organ systems.

In vitro superantigens cause strong T cell activation with subsequent
release of lymphokines like interleukin-2 (IL-2), IL-4, tumor necrosis factor
(TNF) or γ-interferon (γ-IFN) [10–14]. We, therefore, reasoned that superantigen-induced lymphokine release in vivo might be causally involved in
shock. In part, this reasoning is supported by the observation of shock

symptoms in tumor patients treated with high-dose IL-2 [15]. To test our thesis, an animal model was developed in order to study the in vivo consequences of bacterial superantigens. In these studies we used the super-antigen SEB.

Animal Model

It is known that endotoxin (lipopolysaccharide; LPS) released by Gram-negative bacteria cause lethal shock in humans and mice. Macrophages and their product TNF-α/cachectin have been shown to be critically involved in the development of LPS-induced shock, since transfer of LPS-activated macrophages in otherwise untreated mice caused shock and anti-TNF mono-clonal antibody efficiently blocked the development of shock [16–18]. In this model, *D*-galactosamine (*D*-Gal) increased the sensitivity of mice to LPS-mediated lethal shock up to thousandfold [19, 20]. *D*-Gal is a hepatotoxic agent which depletes hepatocytes of UTP by accumulation of UDP-galactos-amine [21]. Subsequently, biosynthesis of macromolecules ceases, thus leading to liver cell damage and cell death at later time points. The precise relationship of impaired liver function and increased sensitivity to shock is, however, not known.

Key to the results described here were our observations that *D*-Gal administration also increased the sensitivity of mice to the lethal shock caused by SEB, a bacterial superantigen. While mice injected subcutaneously with 300 µg SEB showed only mild or no symptoms of disease in the absence of *D*-Gal, upon simultaneous injection of 20 mg *D*-Gal as little as 2 µg SEB per mouse caused lethal shock (table 1). As described for the endotoxin-mediated shock [20], *D*-Gal had to be administered within the first 4 h after or before SEB injection. The first symptoms of disease like piloerection and hypomotility were present 3–4 h after injection of SEB, and after 8 h the animals died. A dose of 20 µg SEB per mouse was sufficient to kill all mice tested, and 2 µg SEB per mouse represented the LD_{50} with some mice dying at later time points (fig. 1). These results were not restricted to one specific mouse strain as other inbred strains of mice showed the same symptoms with slight variation in the dose of SEB used. At the time of death the treated mice were examined histologically. The prominent finding was a necrosis of hepatocytes focused in the center of liver lobules. Taken together, the observations from this animal model demonstrate the development of lethal shock induced by SEB, a product of the Gram-positive bacterium *S. aureus.*

Table 1. Lethal effect of SEB (or LPS) in *D*-Gal-sensitized mice

Treatment			Lethality within 8 h dead/total
D-Gal mg i.p.	SEB μg/mouse	LPS μg i.p.	
20	–	–	0/3[a]
40	–	–	0/3[a]
–	300	–	0/3[a]
–	200	–	0/3[a]
–	20	–	0/3[a]
20	200	–	4/4
20	20	–	4/4
20	2	–	1/4
20	0.2	–	0/4
–	–	100	0/3
–	–	50	0/3
–	–	5	0/3
20	–	10	3/3
20	–	1	3/3
20	–	0.1	1/3

Groups of BALB/c mice received simultaneously *D*-Gal (i.p.) and SEB (hind footpads) or *D*-Gal and LPS (i.p.). Controls received *D*-Gal, SEB or LPS alone.
[a] No apparent signs of illness.

Importance of T Cells for Superantigen-Induced Lethal Shock

As mentioned above, macrophages represent the central cellular element in the pathogenesis of the LPS-induced shock. Because the superantigen SEB activates T cells expressing Vβ8[+] TCRs, about 20–30% of the peripheral T-cell pool becomes activated by SEB. To test whether this SEB-induced T-cell activation is critically involved in the pathogenesis of the SEB-induced lethal shock, two experimental approaches were used. First, immunocompetent mice were treated with cyclosporin (CS), a reagent known to functionally block T-cell activation as well as lymphokine production [22]. Simultaneously, the mice were challenged with SEB and *D*-Gal and compared with a control group receiving only *D*-Gal and SEB. As shown in table 2, CS treatment of mice completely blocked the development of SEB-induced lethal shock, whereas the mice of the control group died, as expected. In a

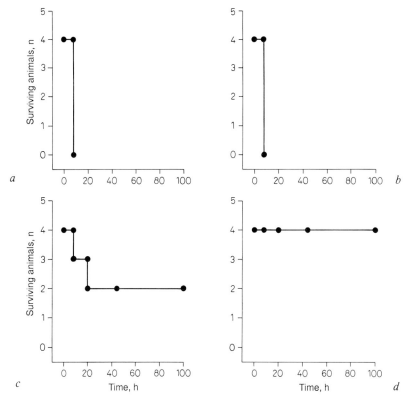

Fig. 1. Lethal effect of SEB. Groups of BALB/c mice (4/group) were simultaneously treated with 20 mg D-Gal (i.p.) and the following titrated amounts of SEB (hind footpads) per mouse: 200 μg *(a)*, 20 μg *(b)*, 2μg *(c)* and 0.2 μg *(d)*.

second set of experiments, immunoincompetent SCID mice, lacking B and T cells, were used. LPS (endotoxin)-provoked shock was readily inducible in SCID mice, indicating the existence of LPS-reactive macrophages. However, administration of SEB failed to induce shock-like symptoms in SCID mice; the animals stayed healthy and showed no signs of disease (table 2). To prove that T cells are responsible for the superantigen-induced shock, T cells from MHC congenic BALB/c mice were transferred into SCID mice, and the repopulated SCID mice were then challenged with SEB. As predicted, the repopulated mice succumbed to the SEB-induced shock. These experiments strongly support our concept of a T-cell-dependent lethal shock induced by the superantigen SEB. At present, experiments are in progress which indicate

Table 2. T-cell-mediated lethal shock syndrome

D-Gal 20 mg i.p.	CS 0.5 mg i.p.	SEB 20 µg/mouse	LPS 10 µg i.p.	Lethality dead/total	
				BALB/c	SCID
−	−	+	−	0/3	n.d.
+	−	+	−	3/3	0/9
+	+	+	−	0/5	n.d.
+	−	+	−	3/3	5/5[a]
+	−	−	+	3/3	3/3

Treatment of mice with more than one agent was done at the same time with the doses indicated.
[a] SCID mice were reconstituted with 15×10^6 BALB/c T cells 6 days prior to injection.

that other bacterial superantigens besides SEB, e.g. TSST-1, are also able to induce lethal shock in this same animal model. In the case of TSST-1, only about 5% of peripheral T cells become activated. To us, this explains the observed slower progression of shock as well as the fatal outcome occurring at later time points. We, therefore, conclude that the number of in vivo activated T cells limits the 'speed' of lethal shock.

Kinetics of T-Cell Activation and Lymphokine Release in vivo

Next, experiments were performed analyzing the in vivo kinetics of the expression of the IL-2 receptor of SEB-reactive Vβ8⁺ T cells. Two hours after injection of SEB into the footpad, T cells of the draining popliteal lymph node were still IL-2-receptor-negative as determined by FACS analysis. However, at later time points (> 4 h), their IL-2 receptor expression became up-regulated, and after 8 h virtually all Vβ8⁺ T cells were IL-2-receptor-positive (fig. 2a). The presence or absence of D-Gal did not influence the kinetics of IL-2 receptor expression.

In addition the serum levels of T-cell-derived lymphokines were evaluated. In contrast to the expression of IL-2 receptors, peak serum levels of IL-2 were recorded already after 2 h, and declined to normal values between 4 and 8 h after injection (fig. 2b). It should be emphasized that peak IL-2 concentra-

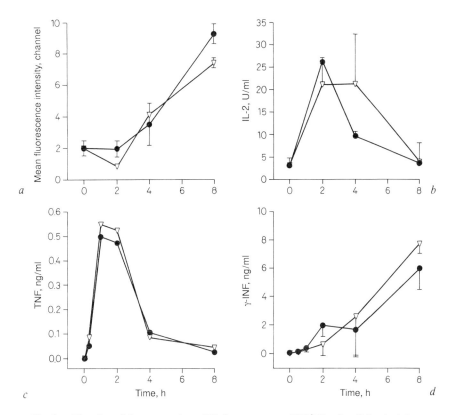

Fig. 2. *a* Kinetics of the expression of IL-2 receptor on Vβ8⁺ T cells of the draining popliteal lymph node. BALB/c mice were either treated with SEB alone (20 μg/footpad; ●) or with *D*-Gal (20 mg i.p.) and SEB (20 μg/footpad; ▽). Each symbol represents 3 individual mice. *b* Kinetics of serum IL-2 levels in mice injected with SEB alone (20 μg/footpad; ●) or with *D*-Gal (20 mg i.p.) and SEB (20 μg/footpad; ▽). Each symbol represents 3 individual mice. IL-2 was measured using the CTLL bioassay. *c* Kinetics of serum TNF levels. At each time point the sera of 2 individual mice, treated with 20 μg SEB and 20 mg *D*-Gal (▽) or 20 μg SEB alone (●), were analyzed. TNF was measured with commercially available ELISA kits. *d* Same experiment as shown in *c*, but serum γ-IFN levels were measured with commercially available ELISA kits.

tions were as high as 20 U/ml of serum. In contrast, serum γ-IFN levels increased slowly throughout the observation period of 8 h (fig. 2d), thus paralleling the kinetics of IL-2 receptor expression.

A central mediator of endotoxin-induced shock is TNF-α/cachectin produced by activated macrophages [18, 23, 24]. To establish whether cytokines of the TNF family are important in superantigen-induced shock, we also measured the serum levels of TNF. Unfortunately, the available murine TNF ELISA kits do not differentiate between TNF-α/cachectin and TNF-β/lymphotoxin. TNF-β/lymphotoxin is exclusively produced by activated T cells [25], whereas TNF-α/cachectin is generated by T cells and macrophages [23]. Already 30 min after SEB application, serum levels of TNF rose and peaked within 1 h (fig. 2c). The serum TNF peak was of short duration, and returned to normal values approximately 4 h after SEB treatment. Taken together, these results indicated that serum TNF levels peaked even faster than those of IL-2, and for both cytokines peak levels were obtained at time points when signs of T-cell activation in terms of IL-2 receptor expression were barely detectable.

Tumor Necrosis Factor Is a Central Mediator of Superantigen-Induced Shock

To further evaluate the relevance of TNF in the SEB-induced shock model, blocking studies with anti-TNF monoclonal antibody were performed. Titrated amounts of anti-TNF monoclonal antibody were injected 1 h prior to a challenge of SEB and D-Gal. The experiments demonstrated that as few as 50 μg anti-TNF monoclonal antibody per mouse were sufficient to protect the challenged mice from lethal shock (table 3). In contrast, anti-IL-2 monoclonal antibody in concentrations of up to 250 μg

Table 3. Anti-TNF monoclonal antibody confers protection

D-Gal 20 mg i.p.	SEB 20 μg/mouse	Anti-TNF mAb μg i.p.	Lethality dead/total
+	+	250	0/2
+	+	50	0/2
+	+	10	2/3
+	+	1	3/3

Groups of BALB/c mice were injected with D-Gal and SEB at the same time. Two hours in advance, titrated amounts of anti-TNF monoclonal antibody (mAb) V1q were administered. Isotype-matched control monoclonal antibodies were without effects (data not shown).

per mouse did not protect. Thus, TNF seems to be a central mediator of superantigen-induced shock.

Desensitization to Superantigen-Induced Shock by Pretreatment with Staphylococcal Enterotoxin B

There is evidence that superantigens not only activate but also tolerize the reactive T cells in vivo and in vitro. In fact, superantigen-reactive human T-cell clones were induced to an anergic state by preincubation with high doses of the respective superantigen in vitro [26]. Furthermore, treatment of mice with SEB apparently causes apoptosis in SEB-reactive T cells 3–4 days later [27]. Therefore, we tested whether pretreatment of mice with SEB alone would tolerize and thus protect them towards a challenge to a lethal dose of SEB and D-Gal at later time points. Indeed, pretreatment of mice with SEB alone 4 h prior to challenge with lethal doses of SEB and D-Gal efficiently blocked the development of lethal shock. This desensitization lasted at least for the next 6 days.

Since we demonstrated that TNF is the critical mediator of SEB-induced shock, serum TNF levels were measured in desensitized mice. Animals were pretreated with SEB, and after 6 h a lethal dose of SEB plus D-Gal was applied. Serum samples were measured during the pretreatment phase and after challenge. Peak TNF values as described above could be recorded after the first injection of SEB. However, upon challenge no TNF production was discernable. Thus, T cells from desensitized mice fail to produce TNF and this failure might be responsible for the protection to superantigen-induced shock.

The above conclusion predicts that the in vivo reactivity pattern of T cells changes over time upon superantigen application. To test this prediction, mice were injected with SEB in the footpad and at later time points the draining popliteal lymph nodes were removed and the T cells restimulated in vitro. T cells exposed to SEB in vivo for less than 1 h responded in vitro with cell proliferation and IL-2 secretion. If, however, the lymph nodes were removed 4 h after injection of SEB, the T cells failed to respond to SEB in vitro as assessed by cell proliferation or IL-2 secretion. However, the same cells vigorously responded to exogenous recombinant IL-2. Thus, the reactivity pattern of T cells changes over time: T lymphocytes stimulated for less than 1 h in vivo can be restimulated in vitro, resulting in proliferation and lymphokine (IL-2) production. T cells stimulated for more than 4 h in vivo fail to respond to SEB and lack the ability to secrete IL-2. However, SEB-induced IL-2 receptor expression appears to be unaltered and the IL-2-receptor-positive T cells respond vigorously to exogenous IL-2.

Open Questions

In contrast to the endotoxin shock in which macrophages represent the central cellular element for the pathogenesis of the shock syndrome, the experiments described here provide compelling evidence that T cells represent the central cellular element in superantigen-induced shock. This is demonstrated by the inability of T-cell-deficient SCID mice to react with shock symptoms upon injection of SEB. This failure could be overcome by reconstitution of SCID mice with immunocompetent T cells. Furthermore, T-cell-derived lymphokines like IL-2, γ-IFN and TNF were detected in the serum of challenged animals, and CS effectively blocked the development of shock. We thus believe that T cells are responsible for the pathogenesis of superantigen-induced shock [28]. However, some aspects of this T-cell-mediated shock are not fully understood. For example, the burst of SEB-mediated IL-2 release was unexpected. Further, we failed so far to detect IL-2 mRNA (by polymerase chain reaction analysis) during the first 4 h after SEB treatment, despite the high serum IL-2 levels observed. Only at later time points (> 4 h) did IL-2 mRNA become detectable with kinetics as described by others [29]. This raises the question of the mechanisms causing the high and immediate serum levels of IL-2 and TNF. It is remarkable that in humans undergoing rejection of kidney transplants anti-CD3 monoclonal antibodies are administered to treat the rejection crisis. Under these circumstances, severe side effects are observed including fever, hypotension and shock, accompanied by systemically increased IL-2 and TNF levels. The kinetics of lymphokine appearance is very similar to the one described here [30–32]. Similar observations were made in mice after anti-CD3 monoclonal antibody injection [33, 34].

TNF-α/cachectin is well known as central mediator of endotoxin-induced shock. Upon injection of LPS, high levels of TNF-α/cachectin can be measured [23, 24] and treated mice are protected from lethal shock by injection with anti-TNF monoclonal antibody [17, 18]. Even though macrophages are involved in the LPS-provoked shock syndrome [16], whereas T cells are responsible for the superantigen-induced shock, it seems that both entities share pathophysiological sequences in which TNF-α/β function as central mediators. We are not yet able to differentiate between TNF-α/cachectin or TNF-β/lymphotoxin at the protein level, because the murine anti-TNF monoclonal antibodies available to us are cross-reactive. Since T cells are the critical cell type in superantigen-induced shock, we anticipate that in the latter system TNF-β/lymphotoxin represents the central mediator. This reasoning does not exclude the possibility that in addition TNF-α/cachectin as produced by macrophages, when activated via SEB-mediated cross-linking of MHC class II molecules, may play a role. To solve this issue, polymerase chain reaction analyses are currently being performed. The data available so far indicate the existence of TNF-β/lymphotoxin message but no TNF-α/cachectin message in cells of lymph nodes draining the side of SEB injection.

The shock model described here shows that T cells become rapidly activated by the superantigen SEB in vivo. However, the ability of T cells to respond with proliferation and IL-2 secretion is confined to a narrow window of time. Thus, within the first hour after SEB application in vivo, T cells can be restimulated in vitro by SEB, but not at later time points. To explain this, we currently entertain several possibilities. First, potentially reactive cells, i.e. Vβ8+ T cells, could have left the draining popliteal lymph node after the first hour of treatment. This rather trivial explanation seems unlikely because the proportion of Vβ8+ T cells does not change over the first 8 h as determined by FACS analyses. In addition, the same cells still respond to exogenous IL-2, indicating that T cells of the draining popliteal lymph node were primed in vivo. Second, the responding T cells could have changed to an anergic state, a phenomenon described for human superantigen-reactive T cell clones [26]. After incubation with high doses of superantigen these clones changed their pattern of reactivity very similar to the pattern described here, in that they could not be restimulated with antigen but responded to exogenously added recombinant IL-2. In addition, the phenotype of the clones changed as evaluated by FACS analyses. The expression of the TCR was down-regulated in contrast to up-regulation of the IL-2 receptor. Experiments are currently in progress to evaluate this possibility in our system. Third, the function of antigen-presenting cells could have changed after administration of SEB in vivo. For example, down-regulation of MHC class II molecules would reduce effective presentation of SEB to T cells, or diminished expression of accessory molecules like LFA-1, ICAM, LFA-3, B7 could also be responsible for the lack of T-cell activation.

Conclusion

The pathophysiologic events characterizing endotoxin-induced lethal shock have been well defined during the last years. In contrast, the understanding of the pathophysiology of the shock syndrome provoked by Gram-positive organisms is yet limited. With the discovery of superantigens produced by staphylococci and streptococci, candidates are available to explain this kind of shock. The strong activation of high numbers of peripheral T cells by superantigens with subsequent release of lymphokines, as described here, appears as one important component in the development of T-cell-mediated shock. In addition, TNF seems to be a critical mediator, both in superantigen- as well as in endotoxin-mediated shock.

The activation of T cells seems to be confined to early time points after injection of SEB. Later, the phenomenon of desensitization can be observed. Further experimentation is needed to define the cellular and molecular mechanisms of desensitization.

References

1   Fleischer B, Schrezenmeier H: T cell stimulation by staphylococcal enterotoxins: Clonally variable response and requirement for major histocompatibility complex class II molecules on accessory or target cells. J Exp Med 1988;167:1697.

2   Mollick JA, Cook RG, Rich RR: Class II MHC molecules are specific receptors for staphylococcus enterotoxin A. Science 1989;244:817.

3   White J, Herman A, Pullen AM, Kubo R, Kappler JW, Marrack P: The V beta-specific superantigen staphylococcal enterotoxin B: Stimulation of mature T cells and clonal deletion in neonatal mice. Cell 1989;56:27.

4   Dellabona P, Peccoud J, Benoist C, Mathis D: T-cell recognition of superantigens: Inside or outside the groove? Cold Spring Harb Symp Quant Biol 1989;1:375.

5   Marrack P, Kappler J: The staphylococcal enterotoxins and their relatives. Science 1990;248:705.

6   Spero L, Jonson-Winger A, Schmidt JJ: Enterotoxins of staphylococci; in Keeler RS, Tu AI: Handbook of Natural Toxins. New York, Dekker, 1988, pp 131–163.

7   Matthes M, Schrezenmeier H, Homfeld J, Fleischer S, Malissen B, Kirchner H, Fleischer B: Clonal analysis of human T cell activation by the Mycoplasma arthritidis mitogen (MAS). Eur J Immunol 1988;18:1733.

8   Crass BA, Bergdoll MS: Toxin involvement in toxic shock syndrome. J Infect Dis 1986;153:918.

9   Choi Y, Lafferty JA, Clements JR, Todd JK, Gelfand EW, Kappler J, Marrack P, Kotzin BL: Selective expansion of T cells expressing V beta 2 in toxic shock syndrome. J Exp Med 1990;172:981.

10 Uchiyama T, Kamagata Y, Wakai M, Yoshioka M, Fujikawa H, Igarashi H: Study of the biological activities of toxic shock syndrome toxin-1. I. Proliferative response

and interleukin 2 production by T cells stimulated with the toxin. Microbiol Immunol 1986;30:469.

11 Jupin C, Anderson S, Damais C, Alouf JE, Parant M: Toxic shock syndrome toxin 1 as an inducer of human tumor necrosis factors and gamma interferon. J Exp Med 1988;167:752.

12 Galelli A, Anderson S, Charlot B, Alouf JE: Induction of murine hemopoietic growth factors by toxic shock syndrome toxin-1. J Immunol 1989;142:2855.

13 Fast DJ, Schlievert PM, Nelson RD: Toxic shock syndrome-associated staphylococcal and streptococcal pyrogenic toxins are potent inducers of tumor necrosis factor production. Infect Immun 1989;57:291.

14 Carlsson R, Sjogren HO: Kinetics of IL-2 and interferon-gamma production, expression of IL-2 receptors, and cell proliferation in human mononuclear cells exposed to staphylococcal enterotoxin A. Cell Immunol 1989;96:175.

15 Rosenberg SA, Lotze MT, Mule JJ: NIH conference: New approaches to the immunotherapy of cancer using interleukin-2. Ann Intern Med 1988;108:853.

16 Freudenberg MA, Keppler D, Galanos C: Requirement for lipopolysaccharide-responsive macrophages in galactosamine-induced sensitization to endotoxin. Infect Immun 1986;51:891.

17 Tracey KJ, Fong Y, Hesse DG, Manogue KR, Lee AT, Kuo GC, Lowry SF, Cerami A: Anti-cachectin/TNF monoclonal antibodies prevent septic shock during lethal bacteraemia. Nature 1987;330:662.

18 Beutler B, Milsark IW, Cerami AC: Passive immunization against cachectin/tumor necrosis factor protects mice from lethal effect of endotoxin. Science 1985;229:869.

19 Lehmann V, Freudenberg MA, Galanos C: Lethal toxicity of lipopolysaccharide and tumor necrosis factor in normal and D-galactosamine-treated mice. J Exp Med 1987;165:657.

20 Galanos C, Freudenberg MA, Reutter W: Galactosamine-induced sensitization of the lethal effects of endotoxin. Proc Natl Acad Sci USA 1979;76:5939.

21 Decker K, Kappler D: Galactosamine hepatitis: Key role of the nucleotide deficiency period in the pathogenesis of cell injury and cell death. Rev Physiol Biochem Pharmacol 1974;71:77.

22 Bunjes D, Hardt C, Röllinghoff M, Wagner H: Cyclosporin A mediates immunosuppression of primary cytotoxic T lymphocytes by impairing release of interleukin 1 and interleukin 2. Eur J Immunol 1981;11:657.

23 Beutler B, Cerami A: The biology of cachectin/TNF – A primary mediator of the host response. Annu Rev Immunol 1989;7:625.

24 Evans GF, Zuckermann SH: Glucocorticoid-dependent and independent mechanisms involved in lipopolysaccharide tolerance. Eur J Immunol 1991;21:1973.

25 Paul NL, Ruddle NH: Lymphotoxin. Annu Rev Immunol 1988;6:407.

26 O'Hehir RE, Lamb JR: Induction of specific clonal anergy in human T lymphocytes by Staphylococcus aureus enterotoxins. Proc Natl Acad Sci USA 1990;87:8884.

27 Kawabe Y, Ochi A: Programmed cell death and extrathymic reduction of Vbeta8+ CD4+ T cells in mice tolerant to Staphylococcus aureus enterotoxin B. Nature 1991;349:245.

28 Marrack P, Blackman M, Kushnir E, Kappler J: The toxicity of staphylococcal enterotoxin B in mice is mediated by T cells. J Exp Med 1990;171:455.

29 Reed JC, Alpers JD, Nowell PC, Hoover RG: Sequential expression of protooncogenes during lectin-stimulated mitogenesis of normal human lymphocytes. Proc Natl Acad Sci USA 1986;83:3982.

30   Chatenoud L, Ferran C, Reuter A, Legendre C, Gevaert Y, Kreis H, Franchimont P,
     Bach JF: Systemic reaction to the anti-T-cell monoclonal antibody OKT3 in relation
     to serum levels of tumor necrosis factor and interferon-gamma. N Engl J Med
     1989;320:1420.
31   Chatenoud L, Ferran C, Legendre C, Thouard I, Merite S, Reuter A, Gevaert Y,
     Kreis H, Franchimont P, Bach JF: In vivo cell activation following OKT3 adminis-
     tration: Systemic cytokine release and modulation by corticosteroids. Transplanta-
     tion 1990;49:697.
32   Gaston RS, Deierhoi MH, Patterson T, Prasthofer E, Julian BA, Barber WH, Laskow
     DA, Diethelm AG, Curtis JJ: OKT3 first-dose reaction: Association with T cell
     subsets and cytokine release. Kidney Int 1991;39:141.
33   Ferran C, Sheehan K, Dy M, Schreiber R, Merite S, Landais P, Noel LH, Grau G,
     Bluestone J, Bach JF: Cytokine-related syndrome following injection of anti-CD3
     monoclonal antibody: Further evidence for transient in vivo T cell activation. Eur J
     Immunol 1990;20:509.
34   Ferran C, Dy M, Merite S, Sheehan K, Schreiber R, Leboulenger F, Landais P,
     Bluestone J, Bach JF, Chatenoud L: Reduction of morbidity and cytokine release in
     anti-CD3 MoAb-treated mice by corticosteroids. Transplantation 1990;50:642.

Dr. med. Thomas Miethke, Institute of Medical Microbiology and Hygiene,
Technical University of Munich, Trogerstrasse 9, D-W–8000 Munich 80 (FRG)

Fleischer B (ed): Biological Significance of Superantigens.
Chem Immunol. Basel, Karger, 1992, vol 55, pp 185–195

Germ-Line-Encoded Recognition of Certain Short Peptide Antigens?

Willi Born[a,b], *Yang-Xin Fu*[a], *Harshan Kalataradi*[a], *Tina Ellis*[a],
Chris Reardon[c], *Kent Heyborne*[d], *Rebecca O'Brien*[a,b,e]

[a] Department of Medicine, National Jewish Center for Immunology and
Respiratory Medicine, Denver; [b] Department of Microbiology and Immunology;
[c] Department of Medicine; [d] Maternal-Fetal Medicine, and
[e] Department of Biochemistry, University of Colorado Health Sciences Center,
Denver, Colo., USA

High frequencies of γδ T lymphocytes (γδ cells) that respond to mycobacterial antigens and 60-kDa heat shock proteins (HSP-60) have been reported in both the mouse and man [1–10]. The expression of a limited set of T-cell receptor (TCR) V genes by these cells, along with diverse TCR junctions, is somewhat reminiscent of 'superantigen' responses of αβ T lymphocytes (αβ cells).

The responses of αβ cells to all bacterial or retroviral superantigens known to date have several features in common that distinguish them from responses to conventional antigens [11]. Firstly, superantigens are V-gene-selective, such that virtually all αβ cells expressing certain V genes are stimulated by a given superantigen. So far, only superantigens that interact with Vβ genes have been unequivocally identified, with one possible exception [12]. Since junctional TCR sequences are unimportant in the recognition of superantigens, responder cell populations are large, and primary antigen responses are easily detectable. Secondly, superantigens are active only as intact molecules [13], unlike conventional antigens which stimulate T cells after being processed into short peptides. This appears to be true for bacterial and viral superantigens alike. A recent analysis of an apparent minor lymphocyte-stimulating gene product, encoded by an endogenous mouse mammary tumor virus, revealed a type II integral membrane protein that binds to TCR and class II major histocompatibility complex (MHC) molecules simultaneously [14]. Thirdly, all superantigens described so far interact with MHC class II molecules.

In this article, we will briefly describe $\gamma\delta$ cell responses to mycobacterial antigens and HSP-60, as well as $\alpha\beta$ cell responses to HSP-60. Using the criteria given above, we will discuss the possibility that $\gamma\delta$ cell HSP-60 reactivity represents a superantigen response.

Response of $\gamma\delta$ Cells to 60-Kilodalton Heat Shock Proteins

Most antigens that stimulate $\alpha\beta$ cells in vivo or in vitro fail to elicit $\gamma\delta$ cell responses [15]. This is surprising in view of the fact that $\alpha\beta$ and $\gamma\delta$ cells resemble each other in many features, in particular the molecular structure of their TCRs. In fact, some of the expressed Vα and Vδ segments are identical and may be derived from the same genes [16]. The lack of reactivity on the part of $\gamma\delta$ cells suggests that their biological role is different from that of $\alpha\beta$ cells, although how they differ is not clear. Among the various possibilities, perhaps the most intriguing is that the specificities of $\gamma\delta$ cells are not directed against diverse foreign antigens, but rather against certain autoantigens [17, 18]. This idea was first proposed after the discovery that the $\gamma\delta$ cells located in various epithelia are composed of remarkably uniform, tissue-specific subsets. For example, epidermal $\gamma\delta$ cells in the mouse are identical to each other with regard to the $\gamma\delta$ TCR they express [18]; similarly uniform $\gamma\delta$ cells colonize the vaginal epithelia or the tongue [19], but express different TCRs specific to these sites. Each of these populations of $\gamma\delta$ cells may therefore recognize only one ligand. It was further suggested that these epithelium-associated $\gamma\delta$ populations might monitor the well-being of the tissues in which they reside, and react to various forms of tissue injury via recognition of abnormally expressed 'self'-molecules peculiar to a given site [17, 18, 20].

However, in some tissues $\gamma\delta$ cells were found to express fairly heterogeneous receptors. Considerable diversity has been documented in $\gamma\delta$ cells from blood, lung, peripheral lymphoid tissues and thymus [21–26]. This was taken to suggest that at least some $\gamma\delta$ cells can recognize diverse ligands and thus probably foreign antigens. If this assumption is correct, different subsets of $\gamma\delta$ cells might have radically different functions.

Nevertheless, the results of experiments with one such fairly diverse subset, the $\gamma\delta$ cells stimulated by mycobacterial antigens, appear to tell a different story. The reactivity of γδ cells with mycobacterial antigens was discovered almost simultaneously in mice and humans [1–3]. A study with hybridomas generated from newborn mouse thymocytes suggested that the relative frequencies of γδ cells reactive with mycobacterial antigens are extraordinarily high, and showed that they express a limited set of TCRs, utilizing a single Vγ gene and a small number of Vδ genes [1, 4]. In this regard, the γδ cell response to mycobacterial antigens resembles superantigen responses. Similar findings have been reported in several studies with human cells [9, 10]. Human peripheral-blood γδ populations can be divided into two major subsets, based on the expression of Vδ1 or Vδ2. Vδ2 is usually found in association with Vγ9, and cells expressing Vδ2/Vγ9 show reactivity with mycobacterial antigens. Although the gene usage is fairly uniform, junctional TCR sequences of these cells are diverse, again suggesting as for a super-antigen response that the junctions may not be involved in the recognition of the mycobacterial ligand. Attempts to isolate and characterize the stimulatory molecule(s) from mycobacteria have not yet been successful.

Mycobacterium-reactive murine γδ cells also respond to mycobacterial HSP-60 [1]. Responder populations reactive with HSP-60 and purified protein derivative or other crude preparations of mycobacterial antigens are entirely overlapping; i.e. the same clones that react with mycobacterial antigens, and no others, also respond to recombinant HSP-60 or at least synthetic peptides representing a portion of HSP-60. This has led us to suggest that the stimulatory activity in the various mycobacterial antigen preparations is probably related to or derived from HSP-60 [1]. To date, this and several alternative possibilities have been examined, but no conclusion has been reached. Although the γδ cell response to HSP-60 was first discovered in thymocytes of newborn mice, it is now clear that γδ cells expressing TCRs with specificity for HSP-60 also exist in adult tissues, including the spleen, epidermis, placenta/decidua and probably liver [27, and unpubl. observations]. High frequencies of HSP-60-reactive γδ cells among randomly generated hybridomas suggest that these cells represent a large portion of the γδ cell repertoire in spleen and thymus [1, 27]. In epidermis and placenta/decidua, frequencies of HSP-60-reactive γδ cells are much lower, and it remains uncertain whether these cells are circulating or resident. HSP-60-reactive responder cells all express Vγ1-J4C4, in association with several Vδ genes. While there is a strict correlation between reactivities of murine γδ cells with crude mycobacterial antigens and HSP-60, it has been difficult to demonstrate responses of human mycobacterium-reactive γδ cells with purified HSP-60 preparations [6]. So far, only two human clones with specificities for HSP-60 have been reported [3, 28], and it remains unclear whether the bulk of human mycobacterium-reactive γδ cells can recognize HSP-60.

Unlike the superantigen responses of αβ cells, HSP-60-reactive mouse γδ cell hybridomas can be stimulated with small synthetic peptide antigens, suggesting that they normally recognize processed and presented protein fragments [29]. Using such peptides, we have identified a putative epitope recognized by these cells, that lies within amino acids 180–196 of *Mycobacterium leprae* HSP-65. More recent unpublished data of our laboratory indicate that the stimulatory activity lies to the amino-terminal half of this sequence, and show that the HSP-60-reactive γδ cells can respond to peptides as short as 9 amino acids. Such antigenic requirements are reminiscent of MHC-restricted αβ cell responses to conventional antigens [30]. Indeed, our most recent data implicate MHC class I or structurally related molecules in the response of γδ cells to HSP-60 [unpubl. observations]. Thus, it appears that HSP-60-derived 9-mer peptides are recognized in a rather conventional fashion, presented by some sort of class I molecule. The identity of the presenter molecule has not yet been determined. It should also be noted that the experiments with synthetic peptides described here cannot safely predict the structure of 'natural' antigens.

The fact that the specificities of all HSP-60-reactive γδ cells studied so far are focused on a single, well-defined antigenic epitope, is rather surprising in view of the relative heterogeneity of their γδ TCRs [27, 31]. Among newborn-thymus-derived cells, we have found productive rearrangements of 4 Vδ genes. In adult spleen, at least 6 different Vδ genes are associated with HSP-60 reactivity. All cells can be stimulated with the same peptide antigen (covering amino acids 180–196 of the *M. leprae* sequence). There are, however, differences in reactivity. Cells that exhibit the strongest response almost invariably express Vδ6, with Vδ6.3 being the most frequently used V gene [27]. This suggests that the variable portion of the TCR δ chain plays some role in ligand recognition by these cells. However, the γ chain appears to be more important, because all HSP-60-reactive cells express the same combination of gene segments, Vγ1-J4C4. Random TCR sequences do not appear to influence recognition of this ligand. Adult-spleen-derived γδ cells, when compared to newborn-thymus-derived cells, exhibit more extensive N-region additions and an increased usage of Dδ1 [27]. Nevertheless, reactivity patterns with HSP-60-derived peptides appear to be unaffected. Hybridomas generated by prestimulation of normal γδ cells with an HSP-60-derived peptide frequently use a Vδ gene rarely found in 'virgin' HSP-60 responders;

the immune cells have, nevertheless, retained some heterogeneity in V-J junctions [Y.-X. Fu, in preparation]. Clearly, more formal studies are needed to assess the role junctional TCR sequences play in the response to HSP-60 with finality.

Although the antigen has not yet been defined in the case of human mycobacterium-reactive γδ cells, the pattern of TCR expression is similar to that of HSP-60-reactive cells in mice. Whereas all responder cells express Vγ9 (usually paired with Vδ2), considerable diversity exists in both γ- and δ-junctional TCR sequences [9, 10]. Presumably naive cells isolated from umbilical cord blood, as compared to peripheral-blood lymphocytes that might have been exposed to antigen, showed similar degrees of junctional variation [10]. Thus, in both the response of murine γδ cells to a small peptide antigen and the response of human γδ cells to mycobacteria, antigen specificities may be essentially germ-line-encoded with little evidence for adaptive changes in TCRs [4, 9, 10]. In the response of a γδ cell clone to an alloantigen, the influence of junctional amino acids has been documented, however [32]. More indirect evidence based on cDNA analyses of TCR mRNA expressed in different tissues also indicates that at least certain γδ junctional sequences may be positively selected in thymus and periphery [33–35]. In this regard, responses of γδ cells to HSP-60 or mycobacteria may differ from responses of γδ cells to other antigens.

Do HSP-60-reactive γδ cells recognize 'self'? We have addressed this question by stimulating HSP-60-reactive hybridomas with peptide antigens representing the equivalent region of the murine HSP-60 homologue [29, 36]. Although the autologous (mouse) sequence was in an initial study found to be less stimulatory, more recent data indicate that murine and mycobacterial HSP-60 peptides are equally potent stimulators for HSP-60-reactive hybridomas. Moreover, not only γδ cells derived from thymus but also peripheral HSP-60-reactive clones generally fail to discriminate between the autologous and the mycobacterium-derived sequence. Reactivity with autologous HSP-60 has also been reported with one human γδ cell clone [37]. Besides the strong responses to murine and mycobacterium-derived peptide antigens, some stimulatory activity was also observed with peptides corresponding to equivalent regions of *Saccharomyces cerevisiae*, *Escherichia coli* and *Chlamydia trachomatis* HSP-60 homologues, whereas the homologous amino acid sequences of 5 other species were not stimulatory. In light of these findings, it seems quite possible that the primary focus of HSP-60-reactive γδ cells is indeed directed towards the 'self'-antigen [1, 29, 30, 36, 37]. The response to mycobacterial HSP-60 could be heteroclitic in nature, and may indicate limitations in the ligand specificity of HSP-60-reactive γδ cells [38–40].

Response of αβ Cells to 60-Kilodalton Heat Shock Proteins

Well before the γδ cell response to HSP-60 was discovered, it was known that members of this family of HSPs are potent antigens for αβ cells [41]. Normal and mycobacterium-immune individuals as well as experimental animals harbor HSP-60-reactive αβ clones in higher frequencies than might be expected based on the relatively low amounts of HSP-60 in antigen preparations such as live or dead mycobacteria, mycobacterial extracts or purified protein derivative that are being used to stimulate immune responses. The high frequencies of responder cells were also surprising in view of the strong evolutionary conservation of HSPs. Because the mycobacterial homologues show approximately 50% amino acid sequence homology with the autologous (human, murine) homologue [42, 43], one might think that the immune system would be tolerant to many of the determinants expressed by these proteins. In light of the strong αβ cell response, it has been suggested that the very structure of the immune system rather favors the recognition of at least some of these evolutionarily conserved antigens. According to this idea, certain autoantigens are immunologically dominant because they represent the immune system's sketchy picture of self, and are each served by a network of antigen-specific and anti-idiotypic T and B cell clones [44]. Thus, if HSP-60 belongs to this group of 'self'-representing autoantigens, preferential responses to similar heterologous HSP-60 homologues might be expected. Indeed, a breakdown of tolerance to HSP-60 may play a role in the development of certain autoimmune diseases such as rheumatoid arthritis [3] and diabetes [45], as well as arthritis induced by adjuvant or streptococcal cell wall in rats [46] and collagen- or pristane-induced arthritis in mice [47].

The strong response of αβ cells to HSP-60 is, however, clearly distinct from αβ cell responses to superantigens. Although relative frequencies of HSP-60-reactive αβ cells are high in an immune response to mycobacteria, it remains difficult to find HSP-60-reactive αβ cells in naive lymphocyte populations. The TCR repertoire of HSP-60-reactive αβ cells has not yet been studied in detail, but the existence of multiple epitopes recognized on these proteins by αβ cells implies stimulation of αβ cells bearing a variety of antigen receptors [48, 49]. One study has suggested a bias towards Vβ11 usage in HSP-60-reactive αβ cells of BALB/c mice [50], but these experiments were carried out after neonatal thymectomy, so that the reactive T cells may not represent the normal T cell repertoire [51]. αβ cell responses to mycobacterial HSP-60 homologues could be elicited with many small synthetic peptide antigens, most of which contained sequences compatible with predicted T cell epitopes [48, 52]. It, thus, seems clear that αβ cells recognize HSP-60 like any other conventional antigen, in processed form and presented by MHC molecules.

A comparison of αβ and γδ cell responses to HSP-60 and HSP-60-derived peptides reveals several differences. Firstly, stimulation experiments with hybridomas generated with naive T cells suggest that the relative frequencies of HSP-60-reactive cells are much higher among γδ cells than αβ cells [30]. Whereas 30–50% of γδ-TCR-bearing hybridomas (generated with newborn mouse thymus or adult spleen cells) showed reactivity with HSP-60 or HSP-60-derived peptide antigens, none of the αβ-TCR-bearing hybridomas that were generated in the same random fusions could be stimulated with these antigens. αβ-TCR-bearing hybridomas with specificities for HSP-60 could, however, be easily generated after prior immunization with HSP-60 [unpubl. data]. Interestingly, several immunization protocols that elicited αβ cell responses to HSP-60 failed to induce γδ cell responses. Only repeated immunization with high doses of the synthetic peptide that had been found stimulatory for the γδ hybridomas, followed by restimulation in vitro, resulted in a detectable γδ cell response [Y.-X. Fu et al., in preparation]. Since this is so far the only example of a γδ cell response in vivo to a defined protein antigen, it remains to be seen whether γδ cells are generally more resistant to in vivo priming than αβ cells.

Germ-Line-Encoded Recognition of Small Processed and Presented Peptide Antigens?

Superantigen responses and γδ T-cell responses to HSP-60 have similar characteristics in terms of the relatively high frequencies of reactive clones and the preferential stimulation of cells expressing certain V genes. However, selective V gene usage has also been observed in αβ T-cell responses to small peptide antigens [53], and is probably a characteristic feature of all T-cell responses to well-defined antigenic determinants. Since HSP-60-reactive γδ cells apparently recognize a single antigenic determinant, their limited V gene usage is not exceptional. The mechanisms of antigen recognition in αβ and γδ cell responses to HSP-60, as compared to αβ cell responses to superantigens, are clearly different. HSP-60 appears to stimulate both types of T cells in the form of processed and presented peptide antigens. What remains is the rather curious focus of a large, somewhat heterogeneous population of γδ cells to a single antigenic determinant. Based on the presently available data, it appears quite likely that this focus is germ-line-encoded. It is possible that the biological function of such a hard-wired response to peptide antigens is different from adaptive, selectable peptide responses [30, 38], and that peptide antigens which are the focus of such hard-wired responses are special antigens. Such special peptide antigens which might also exist for αβ cells could, for example, induce the surface expression of presenter molecules that are recognized by subsets of T cells expressing certain V genes [54]. Alternatively, focused reactivity may be a hallmark of γδ cell function [30] and other subsets of γδ cells may be similarly focused in their ligand specificities, including in particular those populations with uniform TCRs [17]. However, it seems unsatisfactory to conclude that junctional TCR sequences in γδ cells have no function whatsoever, and some of the experimental evidence mentioned above contradicts this possibility. Thus, perhaps γδ cells have dual specificities, and initially respond as members of a subset of cells with related TCRs, and secondarily as affinity-selected clones that recognize a given antigen. Studies on the ligand specificities of γδ cells have only just begun and further surprises may be in store.

References

1   O'Brien RL, Happ MP, Dallas A, Palmer E, Kubo R, Born WK: Stimulation of a major subset of lymphocytes expressing T cell receptor γδ by an antigen derived from Mycobacterium tuberculosis. Cell 1989;57:667–674.

2   Janis EM, Kaufmann SHE, Schwartz RH, Pardoll DM: Activation of γδ T cells in the primary immune response to Mycobacterium tuberculosis. Science 1989;244:713–716.

3   Holoshitz J, Koning F, Coligan JE, De Bruyn J, Strober S: Isolation of CD4⁻ CD8⁻ mycobacteria-reactive T lymphocyte clones from rheumatoid arthritis synovial fluid. Nature 1989;339:226–229.

4   Happ MP, Kubo RT, Palmer E, Born WK, O'Brien RL: Limited receptor repertoire in a mycobacteria-reactive subset of γδ T lymphocytes. Nature 1989;342:696–698.

5   Modlin RL, Pirmez C, Hofman FM, Torigian V, Uyemura K, Rea TH, Bloom BR, Brenner MB: Lymphocytes bearing antigen-specific gamma/delta T-cell receptors in human infectious disease lesions. Nature 1989;339:544–548.

6   Kabelitz D, Bender A, Schondelmaier S, Schoel B, Kaufmann SHE: A large fraction of human peripheral blood γδ⁺ T cells is activated by Mycobacterium tuberculosis but not by its 65 kD heat shock protein. J Exp Med 1990;171:667–679.

7   Fisch P, Malkovsky M, Klein BS, Morrissey LW, Carper SW, Welch WJ, Sondel PM:
    Human Vγ9/Vδ2 T cells recognize a groEL homolog on Daudi Burkitt's lymphoma
    cells. Science 1990;250:1269–1273.
8   De Libero G, Casorati G, Giachino C, Carbonara C, Migone N, Matzinger P,
    Lanzavecchia A: Selection by two powerful antigens may account for the presence of
    the major population of human peripheral γ/δ T cells. J Exp Med 1991;173:1311–
    1322.
9   Ohmen JD, Barnes PF, Uyemura K, Lu S, Grisso CL, Modlin RL: The T cell
    receptors of human γδ T cells reactive to *Mycobacterium tuberculosis* are encoded by
    specific V genes but diverse V-J junctions. J Immunol 1991;147:3353–3359.
10 Panchamoorthy G, McLean J, Modlin RL, Morita CT, Ishikawa S, Brenner MB,
    Band H: A predominance of the T cell receptor of Vγ2/Vδ2 subset in human
    mycobacteria-responsive T cells suggests germline gene encoded recognition.
    J Immunol 1991;147:3360–3369.
11 Herman A, Kappler JW, Marrack P, Pullen AM: Superantigens: Mechanism of T-cell
    stimulation and role in immune responses. Annu Rev Immunol 1991;9:745–772.
12 Rust CJJ, Verreck F, Vietor H, Koning F: Specific recognition of staphylococcal
    enterotoxin A by human T cells bearing receptors with the Vγ9 region. Nature
    1990;346:572–574.
13 Pontzer CH, Russell JK, Johnson HM: Localization of an immune functional site on
    staphylococcal enterotoxin A using the synthetic peptide approach. J Immunol 1989;
    143:280–284.
14 Choi Y, Marrack P, Kappler JW: Structural analysis of a mouse mammary tumor
    virus superantigen. J Exp Med, in press.
15 Raulet DH: Antigens for γ/δ cells. Nature 1989;339:342–343.
16 Raulet DH: The structure, function, and molecular genetics of the γ/δ T cell receptor.
    Annu Rev Immunol 1989;7:175–207.
17 Janeway CA Jr, Jones B, Hayday A: Specificity and function of T cells bearing γδ
    receptors. Immunol Today 1988;9:73–76.
18 Asarnow DM, Kuziel WA, Bonyhadi M, Tigelaar RE, Tucker PW, Allison JP.
    Limited diversity of γδ antigen receptor genes of Thy-1+ dendritic epidermal cells.
    Cell 1988;55:837–847.
19 Itohara S, Farr AG, Lafaille JJ, Bonneville M, Takagaki Y, Haas W, Tonegawa A:
    Homing of a γδ thymocyte subset with homogeneous T-cell receptors to mucosal
    epithelia. Nature 1990;343:754–757.
20 Janeway CA Jr: Frontiers of the immune system. Nature 1988;333:804–806.
21 Augustin A, Kubo RT, Sim G-K: Resident pulmonary lymphocytes expressing the
    γ/δ T-cell receptor. Nature 1989;340:239–241.
22 Itohara S, Nakanishi N, Kanagawa O, Kubo R, Tonegawa S: Monoclonal antibodies
    specific to native murine T-cell receptor γδ: Analysis of γδ T cells during thymic
    ontogeny and in peripheral lymphoid organs. Proc Natl Acad Sci USA 1989;86:
    5094–5098.
23 Cron RQ, Koning F, Maloy WL, Pardoll D, Coligan JE, Bluestone JA: Peripheral
    murine CD3+, CD4−, CD8− T lymphocytes express novel T cell receptor γδ struc-
    tures. J Immunol 1988;141:1074–1082.
24 Elliott JF, Rock EP, Patten PA, Davis MM, Chien H-H: The adult T-cell receptor δ
    chain is diverse and distinct from that of fetal thymocytes. Nature 1988;331:627–
    631.

25 Ezquerra A, Cron RQ, McConnell TJ, Valas RB, Bluestone JA, Coligan JE: T cell receptor δ-gene expression and diversity in the mouse spleen. J Immunol 1990; 145:1311–1317.

26 Groh V, Porcelli S, Fabbi M, Lanier L, Picker LJ, Anderson T, Warnke RA, Bhan AK, Strominger JL, Brenner MB: Human lymphocytes bearing T cell receptor γ/δ are phenotypically diverse and evenly distributed throughout the lymphoid system. J Exp Med 1989;169:1277–1294.

27 O'Brien RL, Fu Y-X, Cranfill R, Dallas A, Reardon C, Lang J, Carding SR, Kubo R, Born W: HSP-60 reactive γδ cells: A large, diversified T lymphocyte subset with highly focused specificity. Proc Natl Acad Sci USA, in press.

28 Haregewoin A, Soman G, Hom RC, Finberg RW: Human γδ T cells respond to mycobacterial heat-shock protein. Nature 1989;340:309–312.

29 Born W, Hall L, Dallas A, Boymel J, Shinnick T, Young D, Brennan P, O'Brien R: Recognition of a peptide antigen by heat shock reactive γδ T lymphocytes. Science 1990;249:67–69.

30 Born W, Happ MP, Dallas A, Reardon C, Kubo R, Shinnick T, Brennan P, O'Brien R: Recognition of heat shock proteins and γδ cell function. Immunol Today 1990; 11:40–43.

31 O'Brien RL, Happ MP, Dallas A, Cranfill R, Hall L, Lang J, Fu Y-X, Kubo R, Born W: Recognition of a single HSP-60 epitope by an entire subset of γδ T lymphocytes. Immunol Rev 1991;121:155–170.

32 Rellahan BL, Bluestone JA, Houlden BA, Cotterman MM, Matis LA: Junctional sequences influence the specificity of γ/δ T cell receptors. J Exp Med 1991;173:503–506.

33 Sim G-K, Augustin A: Dominantly inherited expression of BID, an invariant undiversified T cell receptor δ chain. Cell 1990;61:397–405.

34 Sim G-K, Augustin A: Extrathymic positive selection of γδ T cells: Vγ4Jγ1 rearrangements with 'GxYS' junctions. J Immunol 1991;146:2349–2445.

35 Kyes S, Pao W, Hayday A: Influence of site of expression on the fetal γδ T-cell receptor repertoire. Proc Natl Acad Sci USA 1991;88:7830–7833.

36 Born W, O'Brien RL, Modlin R: Antigen specificity of γδ T lymphocytes. FASEB J 1991;5:2699–2705.

37 Haregewoin A, Singh B, Gupta RS, Finberg RW: A mycobacterial heat shock protein responsive γδ T cell clone also responds to the homologous human heat shock protein: A possible link between infection and autoimmunity. J Infect Dis 1990; 163:156–160.

38 Born W, O'Brien R: The γδ cell response to stress: Unresolved issues and possible significance. Res Immunol 1990;141:595–600.

39 Born W, Harbeck R, O'Brien RL: Possible links between immune system and stress responses: The role of γδ T lymphocytes. Semin Immunol 1991;3:43–48.

40 Born WK, Harshan K, Modlin RL, O'Brien RL: The role of γδ T lymphocytes in infection. Curr Opin Biol 1991;3:455–459.

41 Young RA: Stress proteins and immunology. Annu Rev Immunol 1990;8:401–420.

42 Picketts DJ, Mayanil CSK, Gupta RS: Molecular cloning of a Chinese hamster mitochondrial protein related to the 'chaperonin' family of bacterial and plant proteins. J Biol Chem 1989;264:12001–12008.

43 Jindal S, Dudani AK, Singh B, Harley CB, Gupta RS: Primary structure of a human mitochondrial protein homologous to the bacterial and plant chaperonins and to the 65-kilodalton mycobacterial antigen. Mol Cell Biol 1989;9:2279–2283.

44 Cohen IR, Young DB: Autoimmunity, microbial immunity and the immunological homunculus. Immunol Today 1991;12:105–110.
45 Elias D, Markovits D, Reshef T, Van Der Zee R, Cohen IR: Induction and therapy of autoimmune diabetes in the non-obese diabetic (NOD/Lt) mouse by a 65-kDa heat shock protein. Proc Natl Acad Sci USA 1990;87:1576–1580.
46 Van Eden W: Heat-shock proteins in autoimmune arthritis: A critical contribution based on the adjuvant arthritis model. APMIS 1990;98:383–394.
47 Van Eden W: Heat-shock proteins as immunogenic bacterial antigens with the potential to induce and regulate autoimmune arthritis. Immunol Rev 1991;121:5–28.
48 Brett SJ, Lamb JR, Cox JH, Rothbard JB, Mehlert A, Ivanyi J: Differential pattern of T cell recognition of the 65-kDa mycobacterial antigen following immunization with the whole protein or peptides. Eur J Immunol 1989;19:1303–1310.
49 Lamb JR, Ivanyi J, Rees ADM, Rothbard JB, Howland K, Young RA, Young DB: Mapping of T cell epitopes using recombinant antigens and synthetic peptides. EMBO J 1987;6:1245–1249.
50 Iwasaki A, Yoshikai Y, Yuuki H, Takimoto H, Nomoto K: Self-reactive T cells are activated by the 65-kDa mycobacterial heat-shock protein in neonatally thymectomized mice. Eur J Immunol 1991;21:597–603.
51 Bogue M, Candéias S, Benoist C, Mathis D: A special repertoire of α:β T cells in neonatal mice. EMBO J 1991;10:3647–3654.
52 Rothbard JB, Taylor WR: A sequence pattern common to T cell epitopes. EMBO J 1988;7:93–100.
53 Moss PAH, Moots RJ, Rosenberg WMC, Rowland-Jones SJ, Bodmer HC, McMichael AJ, Bell JI: Extensive conservation of α and β chains of the human T-cell antigen receptor recognizing HLA-A2 and influenza A matrix peptide. Proc Natl Acad Sci USA 1991;88:8987–8990.
54 Imani F, Soloski MJ: Heat shock proteins can regulate expression of the Tla region-encoded class Ib molecule Qa-1. Proc Natl Acad Sci USA 1991;88:10475–10479.

Willi Born, PhD, Department of Medicine, National Jewish Center for Immunology and Respiratory Medicine, Denver, CO 80206 (USA)

Subject Index